Printed in the USA.

Book illustrations and cover design by Davor Ratkovic

Contents

Full of Hot Air

A rooster, a duck, and a sheep all get into a hot air balloon. While this sounds like the start of a bad joke, it actually happened! In 1783, the first hot air balloon was launched with a rooster, a duck, and a sheep as the passengers.

Why a rooster, a duck, and a sheep? Well, would you want to volunteer to go up in the first hot air balloon without knowing if it would be successful or not? It's always been pretty common for animals to take a first uncertain trip on something new, like how monkeys and dogs went up into space before people.

Jean-François Pilâtre de Rozier, a French scientist credited with being the first balloonist, wanted to create a new way for people to see the world. His balloon used hot air to assist with the lift, but also had a compartment for gas that was lighter than air, likely hydrogen or helium. This first flight with our three animal heroes, lasted 15 minutes.

After that, Jean-Francois got brave. He teamed up with François Laurent d'Arlandes and the two went up in the hot air balloon that was built by the Montgolfier brothers. It's said that King Louis XVI initially wanted a criminal to be the first human to go up in the balloon, in case it crashed. But Jean-Francois really wanted the opportunity so he petitioned the King and won.

And then, ballooning took off...literally! In 1784, Vincenzo Lunardi launched his balloon in London in front of a crowd of 200,000 people. In the basket for the ride with him were three different animals; a dog, a cat, and a caged pigeon. His balloon flew for 24 miles and he helped build the intrigue and popularity of ballooning.

The next year, Frenchman, Jean-Pierre Blanchard, and an American, John Jeffries, took on the English Channel, soaring 21 miles in their balloon over the channel to deliver a letter. This was considered a big step in long distance ballooning. Unfortunately, this same year, our original balloonist, Jean-François Pilâtre de Rozier, died when his balloon exploded in an attempt at crossing the channel.

The next great feat was in 1793 when Jean-Pierre Blanchard sailed to America and made the first balloon ride in North America.

Ballooning became popular, even being used to help with observation during times of war. Of course, now people mostly think of romantic balloon rides, but balloon enthusiasts enjoy balloon races, and challenges as well. People have even sailed balloons all the way around the world. And to think that it all started with a bird, a duck, and a sheep.

London Bridge Is Where?

Bridges are pretty spectacular feats of engineering. Think of the amount of cars, semi trucks, and people that travel across bridges every day. That's a crazy amount of weight that they must hold.

Have you heard the nursery rhyme, London Bridge is Falling Down? The bridge across the Thames River in London DID fall down, several times in fact. It had to be rebuilt each time. The bridge was first built in 43 A.D. Yes, that's almost 2,000 years ago. This first bridge was said to be planks of wood laid across ferry or pontoon boats.

Over the course of many years, the bridge had several tragic endings. In 1014, the Vikings sailed up the river and attached cables to the bridge, using their ships to pull the whole bridge down. It was rebuilt, this time from stone, but with a lot of wooden infrastructure. Over the next 500 years, several fires would cause damage and the bridge had to be repaired several times.

In 1831 a new bridge was opened. This bridge was 928 feet long, 49 feet wide and the base of the bridge consisted of five huge granite arches. London traffic began to use the bridge immediately and by 1896, eight thousand people crossed it everyday. It was even widened to 65 feet to help accommodate all of the traffic.

Unfortunately, the growth of London was too much for the bridge. In 1962, engineers determined that London Bridge was indeed falling down. It was actually sinking down into the river because of the huge amount of weight on it every day. London decided that it was time to build a new bridge. But what would they do with this old bridge that held so much history?

They decided to sell it to the highest bidder by auctioning it off. What? Can you actually sell an old bridge? It turns out you can, but it will cost a pretty penny. So who bought this giant bridge and what did they do with it?

In the 1960's Robert P. McCullock was building a city. Located on Lake Havasu in Arizona, Robert needed a bridge to reach an island in the lake. He purchased the London Bridge for $2,460,000! He then spent $7 million moving the bridge from London to Arizona.

So how did that all work? It took three years. First, the old bridge had to be disassembled. The City of London was smart about this. They built the new bridge right over the top of the old bridge. They were actually able to arrange the construction so that they never lost one day of traffic on the bridge.

As each piece was taken from the old bridge, it was labeled with numbers. This would help when it was rebuilt in Arizona. When the whole bridge was disassembled, it was loaded on boats and sailed 10,000 miles to California. From there it was taken by truck to Lake Havasu City.

In 1968, the Lord Mayor of London, Sir Gilbert Inglefield, laid the cornerstone and the reconstruction of London Bridge in Arizona began. Robert Beresford was the civil engineer tasked with putting the bridge back together. He used the original blueprints from 1831 and used the numbers that had been placed on each piece when it was taken apart. During the reconstruction, the builders used sand mounds underneath each arch to help with support. When the bridge was complete, the sand was taken out, the channel got dredged and the water from the river was diverted underneath the bridge and then into Lake Havasu.

Three years later, construction was complete and the London Bridge was dedicated with many American and British officials present, as well as a crowd of 50,000 people. And here's a fun fact. The sale of land to build houses and businesses in Lake Havasu was so great, that Robert McCullock was able to make back all of the money he had spent on the bridge! The London Bridge still stands in Arizona and doesn't seem to be in any danger of falling down.

The 9 Year Old with a Dagger Made from a Meteor

When I was a kid, I thought that knives were the coolest. I had a survival knife that had a sheath with a sharpening stone. But the really cool thing about it was that you could unscrew the end of the knife handle and pull out all kinds of little survival tools like a fishing line, a hook for catching the fish, a rope saw, and so on. Oh, and the end of the handle was a compass. But my knife was nothing like the one another kid had. This kid was the leader of a country though. And he lived over 3,000 years ago.

You may have heard of King Tut, which is short for Tutankhamen. King Tut became pharaoh when he was just 9 years old and ruled for 10 years before he died of malaria at age 19. He really wasn't a very important Egyptian pharaoh except for the fact that his tomb was so well preserved and has allowed scientists to learn a lot about him and life back then. His tomb was full of all kinds of fascinating things. But one of them in particular, seems truly special.

I'm talking about a knife in a golden sheath, just over a foot long. When Tut's tomb was first discovered, this knife was probably one of the least impressive things found in there. There was the burial mask and coffin that were made out of gold, as well as chariots, thrones, statues, and many other incredible things.

But just a few years ago, an amazing discovery was made about the king's knife. Are you ready for this? It had once flown through space!

It wasn't flying through space as a knife, of course. The iron that the knife was made of came from a meteorite that had crashed into earth...an actual space rock. That's pretty amazing!

We know this thanks to X-ray technology. It's actually called X-ray fluorescence spectrometry. This tells us what something is made of. And this study shows that King Tut's knife is made from the kind of iron that only falls from space. And this makes sense because the knife is from the Bronze Age which was before the Iron Age. In ancient Egypt, people had not yet learned how to make their own iron. That came later. Before the Iron Age when people figured out how to make it, iron was extremely rare which made it more valuable than gold. Most iron artifacts from the Bronze Age came from meteorites.

We only know this now, because X-ray fluorescence spectrometry technology allows scientists to test ancient artifacts without damaging them. That's been a fantastic innovation and allows scientists to find out so much more about ancient times.

It's probably a good thing I didn't have a knife made from a meteor when I was a kid. I'd probably have lost it and been in LOTS of trouble!

Was the Father of Modern Vaccines a Slave from the 1700s?

Science is full of spectacular stories and this is one of them. Let me introduce you to Onesimus, a slave who saved the lives of hundreds of people from the Boston smallpox outbreak of 1721.

Because Onesimus lived as a slave, his birthday and real name are unkown. All we do know is that he was brought to America from Africa in 1706 in the Atlantic slave trade. Historians think he may have been from either Libya or Ghana. In December of that year, Onesimus was in Massachusetts where he was bought as a slave and given as a gift to a Puritan reverend, Cotton Mather.

While he was working for Cotton, Onesimus explained to him the concept of inoculation. Inoculation is very similar to vaccination. Onesimus had been inoculated against smallpox back in Africa. This process involved cutting the skin and dropping one drop of the smallpox virus into the cut. This is basically what a modern day vaccine does. It introduces a weakened version of a virus into our bloodstream so that our bodies begin to build up natural resistance to the actual virus.

It turned out that Onesimus knew a lot more about medicine than most doctors in America at that time.

Inoculation against disease had been practiced for a long time in Africa. It had started almost 3,000 years earlier in China and spread to Africa from there. But these practices were basically unknown to Americans.

A few years after Onesimus had told Cotton about his smallpox inoculation, the disease arrived in Boston. The year was 1721 and a boat arrived in the harbor with a sailor who had smallpox. It spread quickly and was very deadly. Nearly half of all Bostonians got the disease and nearly 1 in every 6 people died from it.

Cotton, to his credit, began to champion the benefits and need for inoculation. Of those who did get inoculated, only 6 died. This was a great victory. But it took a lot of work to convince people as most leaders in Boston thought that inoculation would only spread the disease more. They were also suspicious of any idea that came from a slave.

Of course, Cotton got all of the credit for saving these people's lives with Onesimus' idea. But historians have done a lot to shine light on the contribution that was made by Onesimus and how his knowledge is actually what eventually led to the creation of vaccines in America. In 1796, the first smallpox vaccine was finally invented and used effectively. This happened because Onesimus shared the science of inoculation in a way that was able to convince Cotton that it worked. Onesimus was eventually able to buy his freedom from Cotton and lived his final years as a free man. And in 2016, he was finally recognized as one of the greatest

Bostonians of all time by Boston magazine. Today, smallpox is no longer a threat to anyone. It has gone from being one of the world's deadliest diseases, to completely eradicated. Thank you, Onesimus!

The Mystery of Skeleton Lake

If you enjoy hiking and adventure, you might want to consider trekking up to Skeleton Lake in the Himalayan Mountains. Skeleton Lake? Hmmm... on second thought, that place sounds kinda creepy. What's the deal there?

That's the crazy thing. Scientists, researchers, and archaeologists are more confused than ever on the mystery of Skeleton Lake. Here's what they do know.

High up in the Indian Himalayan Mountains, nearly 16,000 feet up, is a small lake about the size of a football field. The lake is called Roopkund. Back in 1942, an Indian forest official found the lake and hundreds of human skeletons scattered in and around it. That's right, human skeletons! He reported that between 300-800 people had once met their death at this mysterious lake.

Creepy, right? It gets even more interesting. In the 1950s, researchers initially believed that the bodies had to be the remains of Tibetan journeymen or perhaps soldiers. It seemed logical that this group had fallen ill due to an epidemic or perhaps died from exposure to the harsh climate. But by 2004, forensic analysis had become more advanced and scientists were able to determine that the bones were from both men and women, all of various ages. There were several of the remains that had unhealed skull injuries. Yet, there were no weapons at all found at the sight. They did find ancient musical instruments though. Hmmmm.....

So putting together all this new evidence, it was decided that the skeletons were likely from a group of Indian pilgrims making a Hindu pilgrimage. This journey happens once every twelve years and is called Nanda Devi Raj Jat Yatra. Roopkund would have been along the way to the final destination of Homkund. The theory was that the pilgrims were caught in an incredibly rare giant hail storm that produced hail so large that it injured and killed the travelers.

This theory held up until continued scientific advances allowed the first ancient human genome to be sequenced. Then the mystery was opened all over again. Skeleton samples from 38 of the human remains were gathered and sent to labs all over the world for genomic and biomolecular analysis. This meant they could break the bones down to find out more about who the people were and where they came from. The results were astounding!

It turns out that the skeletons that were studied came from three distinctly different groups and locations. It was likely not a single event, but a span of 1000 years in which these three different groups died. The largest percentage of people came from southern Asia and likely died from one or several events that took place between the 7th-10th century. The next group was from much later in the 19th century (1800s). It was a group of Mediterranean people, likely from Greece or Crete. This is very interesting to archaeologists who are trying to figure out why this group was so far away from their homeland and why no one had heard of them before. The smallest group, also from the 19th century, was likely from southeast Asia.

So how in the world did the skeletons of all these people who lived hundreds of years apart and were from different parts of the world end up in the same little lake in the Himalayan Mountains?

Scientists working on data analysis say that there are now more questions than answers to the mystery of what really happened at Roopkund, or Skeleton Lake. It seems that with each scientific advance, new theories are suggested. Will the mystery ever be solved? I have no idea, but I do know that I'll probably stay clear of Skeleton Lake just to make sure I don't end up as a lake skeleton myself.

Dinosaur Bone Battles

Do you know what a rivalry is? It's a competitiveness between people or teams looking to be the best at something. Often the first thing that would pop into our mind would be a sports team rivalry. But the major rivalry in the late 1800s didn't have anything to do with baseball or football. It had to do with bones.

Yep, that's right. This era was huge for discovering dinosaur bones and two of the major paleontologists at the time didn't like each other. In fact, they couldn't stand each other and tried everything they could to discredit and outdo the other.

Edward Drinker Cope was born into a wealthy family and went on to become a professor of zoology at Haverford University in Pennsylvania. Othniel Charles Marsh grew up poor, but he had a rich uncle, George Peabody. When his uncle built the Peabody Museum of Natural History at Yale University, he appointed Othniel as the head of the museum. Upon George's death, Othniel inherited his fortune, making him a very wealthy man.

The two started off as friends, meeting in Berlin and discussing science and fossils. They later exchanged letters and even named species after each other. While working on a dig site in New Jersey, Edward invited Othniel to come and dig with him. This is when things started to go bad. Othniel apparently bribed the pit operators to funnel more of the fossils to him instead of Edward. Very sneaky!

Of course this made Edward mad, so he retaliated by slandering Othniel in newspaper articles and starting to dig for fossils in Kansas in the exact location Othniel was digging. During another incident, Edward was constructing a full skeleton and put the head on the tail instead of the neck. Othniel wouldn't let it slide and let all the newspapers know about Edward's mistake.

During this time, there were many fossils being found out west. So both Othniel and Edward gathered teams and headed to the great frontier. This is when the bone wars really heated up.

Edward and Othniel were both intent on being the best. They each wanted to find the most bones and identify the most new species. And they used some troubling ways to try and achieve these goals. They spied on each other, stole each other's workers, and even destroyed fossils to prevent the other from finding them. They bribed people to keep areas with fossils a secret. They were constantly writing bad things about each other in newspapers and scientific journals.

In the end, Othniel discovered 80 new species and Edward uncovered 56. Just before his death, Edward issued Othniel one more challenge. He vowed to donate his brain to science and urged Othniel to do the same. He wanted to compare brain sizes. In the late 1800's it was thought that the larger the brain, the smarter a person was. Edward wanted to prove he was smarter than Othniel, even in his death. Othniel refused the challenge.

Even with all the sabotage, an incredible amount of new dinosaur species were discovered, including dinosaurs that you have heard of today. Stegosaurus, Allosaurus, and Apatosaurus were all discovered during the Bone Wars. And because of all the scandals in the newspapers, people became more interested in fossils and dinosaurs. I guess you could say that all of mankind actually benefited from the silliness of the Bone Wars.

Nikola Tesla's Death Ray and How He Fell in Love with a Pigeon

You walk into a room and turn on a light. Ta da! It's a pretty common occurrence these days. Everyone associates light and electricity with Thomas Edison, but there were a few other important players when electricity was first being introduced into homes.

Meet Nikola Tesla, the inventor of AC or alternating current. This type of current is responsible for powering nearly all houses and businesses because it can be used over longer distances. It is also much easier and safer to use than DC or direct current.

Nikola was a Serbian-American scientist and inventor. He had a fascination with electricity and was considered to be a bit of a genius. But like many geniuses, he also had some weird habits and ideas. Let's dig in!

Nikola had an obsession with the number three. He would wash his hands three times and walk around a building three times before entering it. When he stayed in a hotel, he made sure that the room number was divisible by the number three. He actually lived the last ten years of his life at the Hotel New Yorker, room 3327.

He wore white gloves every night to dinner and calculated the volume of food on his fork before eating each bite. For these reasons, he prefered dining alone. Actually, Nikola preferred being alone in general. He never married and this never bothered him. He once stated, "I do not think you can name many great inventions that have been made by married men." He also detested pearls and women's jewelry. It is said that when his secretary wore a pearl necklace, he made her go home for the day.

He loved pigeons though and was known for finding ailing pigeons and nursing them back to health in his hotel room home. Some of the neighbors complained about the noise and the smell, but that didn't stop him. He found one bird and spent $2000 on her care when she got sick. He claimed he loved her and that as long as he had her there was a purpose to his life.

All of these oddities don't obscure the fact that Nikola was incredibly brilliant. He even worked for Thomas Edison, although not for very long. Thomas recognized how smart and dedicated Nikola was. He told Nikola that he would pay him $50,000 to improve upon Thomas' DC system. Nikola toiled away at it, but when he presented it to Thomas and asked for payment, Thomas only laughed and said Nikola didn't understand American humor. Rude!

Another invention that Nikola was working on before he died was a death ray. Yep, like you would see in a crazy sci-fi movie. His death ray would use an intense beam of particles or radiation to destroy military targets up to 250 miles in the air. But unlike its name, the "death ray," Nikola actually hoped it would be used as an anti-war device. Nikola hated war and hoped countries would be able to use the ray for protection. The death ray was never actually created, but its mystery and possibility was talked about by the governments of many countries worldwide.

As researchers and other scientists have studied Nikola's life, they have realized that he probably suffered from Obsessive Compulsive Disorder and maybe some other mental health issues. That would explain why he had so many peculiar habits. Unfortunately, back in those days these issues weren't as widely recognized and Nikola never got any help. But he was an incredibly intelligent man who made many advances in science and electricity. He was very forward thinking and helped make many advances that make our lives easier today.

Talking Trees?

Not all scientific research has to do with modern technology. Some scientists are studying something that has been around since the beginning of the earth, yet they are still making new discoveries about how they grow, survive, and thrive.

I'm talking about trees. These giants of the forest didn't just appear magically overnight. Instead, some large trees have been around for hundreds of years. Imagine the stories they could tell! If only they could communicate. It might surprise you to know that scientists are starting to think that trees CAN communicate. They actually "talk" with each other in really cool ways to help each other.

Originally, it was thought that trees existed but were completely independent of each other. A seed sprouted, it had its own roots system, and it had to fight for sunlight in order to grow. Over time the tree's roots grew and so did the branches, limbs, and trunk. Trees were thought of as loners, competing for the nutrients to survive, the strong winning and growing tall, while the weaker trees lost the fight.

Some of this is true, but scientists are discovering something that's pretty fascinating. Perhaps trees in a forest are all interconnected through their roots system. They can share food and information. Some researchers jokingly call it the Wood Wide Web.

It is now believed that trees of the same species work together to help each other survive. They may form alliances with other species that benefit both groups of trees. When we are out in the forest, our eyes are generally drawn up to the soaring tops of the trees, but really all the action is taking place below the ground.

All trees are connected by an underground fungal network. This is where trees get their food and water, and recent findings show that they may be sharing these resources, as well as communicating. They can send distress signals to warn other trees about danger, such as insect infestations, fungal problems, disease, and even droughts. The other trees can then change their behavior to help them adapt. Small saplings that don't yet reach the sunlight are actually fed through this underground system, helping them to survive and grow until they are tall enough to reach the light on their own. In a way, it is almost like how parents feed their young.

Trees also have the ability to communicate through the air, by use of pheromones and other scent signals. In the hot Sahara desert, giraffes feast on the leaves of the acacia tree. When they start eating, the tree detects the damage and emits ethylene gas. This is a distress signal that nearby acacia trees receive. As a response, the nearby trees start pumping tannins into their leaves. These tannins, eaten in a large enough quantity, have the ability to kill a giraffe. So the giraffes stay on the move, never feasting in one place too long, thus protecting the acacia trees.

Research has also shown that trees can tell a difference in deer biting the bark of a tree, versus natural or human damage. Something in the deer saliva can trigger a response that makes the leaves and bark taste bad to the deer. With human or other natural trauma, the tree just emits different chemicals to heal the damaged spot.

So the next time you are in the forest, stop and marvel, not only at the great size of the trees, but also the incredible network of trees communicating, sharing nutrients, and helping each other to adapt that has allowed these forests to grow. It's pretty cool to think about trees helping each other survive.

The Strange Truth about Wasp Faces

You recognize your friends and your family when they walk into a room, right? But what about wasps? Those darn stinging buggers all look the same, and are quite terrifying! So can they recognize each other in order to work together?

It turns out they can. Sometimes scientists spend many hours working to prove or disprove a theory and sometimes they stumble across new discoveries while performing other tasks. Such is the case with Elizabeth Tibbets and her interesting discovery about wasps.

Elizabeth was a college grad student studying wasps. She was studying the hierarchy within wasp colonies. That's who are the leaders, who are the workers, and how they interact with one another.

In order to do this, she would paint each different wasp with a different color spot using model airplane paint. That way she could tell them apart. She would video their interactions and later watch and analyze her findings.

One day she realized she had forgotten to paint a few of the wasps. Darn, all that time wasted! But it turned out that she hadn't wasted any time at all and may have stumbled across an interesting discovery. Elizabeth realized that even without the paint, she could tell the wasps apart. And this made her wonder if the wasps could tell each other apart, too.

To an experienced scientist, this idea might seem preposterous. Bugs don't recognize each other's faces! But Elizabeth was just a grad student and she was curious. So she performed some research and found some interesting results that disproved what more experienced scientists thought.

Wasp faces are actually quite different. They have varying lengths of antennae, different colors, spots, and markings that make them all unique. And Elizabeth's research proved that not only can wasps tell each other apart, they use facial recognition just like humans! This is helpful as they perform tasks and carry on with life in their colonies.

So while it's super cool to learn that wasps can recognize their friends, the real moral of the story is that Elizabeth Tibbets was inquisitive enough to perform her own research. So be like Elizabeth and stay curious. Maybe you will discover something super cool, too!

Can Junk Food Make You Go Blind? Don't Let Your Parents Read This!

"Eat your vegetables!" I'll bet you've heard this a few times in your life, right? It seems like parents are always nagging us to eat vegetables and other healthy foods. But it turns out they may be right. Let's hear this story about a boy who didn't eat his vegetables, and the devastating outcome it had on his life.

There have been many documented cases of people who have lost sight or hearing because they live in areas where food is scarce. The lack of nutrients has an adverse affect on their body and they can temporarily or even permanently lose sight or hearing. This occurs in areas of famine or sudden food shortages.

But a seventeen year old boy in England was recently diagnosed with nutritional optic neuropathy, the loss of sight and reduced hearing capabilities. And it wasn't because of a lack of food. It was because of poor food choices!

The boy admitted to his doctors that since he was in elementary school, he had pretty much only had a diet of french fries, white bread, potato chips, and sausages. While this might sound delicious, it is very lacking in vitamins, calcium, and magnesium. And it turns out that these are pretty vital to our health!

When he was 14, he went to the doctor complaining of being tired. One year later he reported having a harder time than normal hearing and by the time he was 17, he was legally blind. And while some cases of nutritional optic neuropathy can be reversed or slowed, in this boy's circumstances, he was diagnosed too late and will likely be dealing with this the rest of his life.

While cases like this are rare, it's a BIG wake up call that what we feed our bodies does matter! It's important to eat a variety of foods and to be sure that we are eating foods that have nutritional value and are full of healthy vitamins and minerals. Turns out mom was right, those veggies are pretty important. Now, if you don't mind, I'll just be over here eating this broccoli.

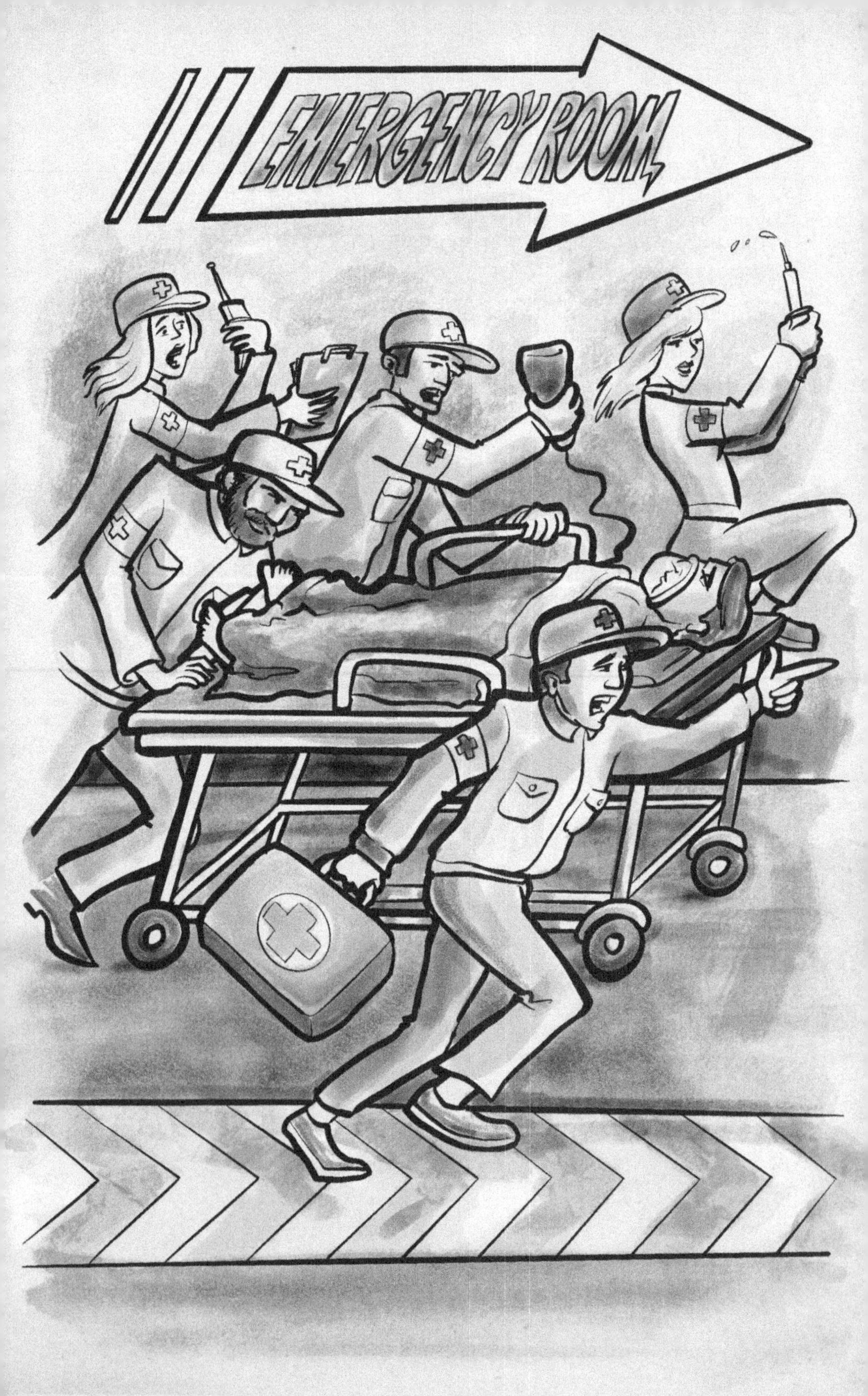
EMERGENCY ROOM

Ferrari Pit Crew Gives Hospital Surgery Staff a Lesson on Efficiency

Great Ormond Street Hospital had a problem. As the largest children's hospital in England, about 100,000 patients are treated here each year, many of whom have various heart diseases. Of course, the stakes are high anytime children are being operated on, but in the early 2000's the hospital saw a big rise in the number of deaths of their young patients.

This was terrible! The doctors and hospital staff certainly thought so and they began to look at what the problem could be. They investigated surgical procedures and things like over-tiredness of the staff. But the main source of problems seemed to always stem from the same place: the hand-off of the patient from the surgery team to the Intensive Care Unit.

When a person is in the hospital, they will be operated on by a group of surgeons and nurses and then transferred to a care team in the Intensive Care Unit or ICU, for recovery. This hand-off seemed to be the place where the most mistakes were made. And sometimes, even just a few very small mistakes could add up to big problems when someone's life is on the line.

Two doctors, Dr. Goldman and Dr. Elliot, were watching a car race on television one day after a hard day of performing surgeries. They were both big automobile racing fans and as they sat and watched the pit crew for the racecars, they couldn't help noticing the similarities to their hospital job. The pit crew had to perform their job quickly and efficiently, with minimal error, to get the car back into the race. Just like in the surgical hand-off, certain tasks needed to be performed to get the patient transferred to the ICU.

They decided to investigate this further and see if it could help their situation. They were able to meet with Ferrari's Formula One pit crew and technical director, Nigel Stepney, to go over the hospital procedures.

Nigel was surprised at the number of mistakes and lack of attention to detail during the patient transfers. He pointed out that there wasn't a single person in charge of the procedure. In racing, a pit crew has a "lollipop man" who directs the flow and speed of the procedures to get the car back out onto the track. While the pit crew is performing their duties, each team member has an assigned task. And these tasks are performed in complete silence. However, at the hospital, staff members were often switching the jobs they did and there was a lot of chatter between doctors and nurses.

Dr. Goldman and Dr. Elliot took these critiques seriously and went back to the hospital ready to hash out a better way to perform. They applied what they had learned from the racing crews and streamlined the way that patients were transferred.

It seemed to work! A study done a year later showed that technical errors were down 42% and errors caused by misinformation during the hand-off were down 49%. The hospital staff had been trained to perform their tasks, but also prepared in case of an emergency. The efficiency and ease of the transfers made everyone, especially the patients, more comfortable and safe. Inspiration can come from anywhere.

The Famous Oceanographer Who Wanted to Be in the Air

In this book we talk about famous researchers who explore space and land, but what about all the cool stuff going on in our oceans? There's got to be some pretty fascinating characters who have explored there, right?

You bet there are! The ocean is full of unbelievable discoveries. And just like outer space, it can be dangerous, full of mystery, and in some places...just downright hard to explore.

Jacque Cousteau was one of those famous first explorers. He loved exploring the oceans and researching every type of living thing below the water's surface. And luckily for everyone else, he liked to share his adventures. But Jacque didn't always want to be below the water. When he was a boy he wanted to be a pilot in the French Navy. He was actually in aviation training when he was in a horrible car accident and was severely injured. Resuming his flight training seemed nearly impossible. Jacque started swimming in the Mediterranean Sea as part of his rehab after the crash. He made a pair of goggles and discovered a whole new world below the ocean's surface. He had found his life's work.

Jacque began to explore underwater anywhere he could. He did underwater research for the French Navy. He and a friend took an underwater camera and filmed a documentary about the ocean. This was exciting to him. He loved being able to share the ocean with other people. He shared his idea of making more movies with a friend, Thomas Loel Guinness. Guinness was enthusiastic about the project and wanted to help. He was also quite wealthy. He bought a car ferry (a large boat) and leased it to Jacque for 1 franc (about $1) per year. Sweet deal!

With the new boat that he and his wife named the Calypso, Jacque and his family took to the seas. He wrote books about his findings. He was the first person to suggest that whales and porpoises use echolocation to be able to navigate through the water. He co-invented the aqualung, which made it possible to breathe underwater for long periods of time and to go deeper into the ocean than before. Have you ever heard of scuba diving? Thanks to the aqualung, many people today enjoy scuba diving. He also invented many of the underwater cameras and lighting systems that are still used today.

Did you know that Jacque won 3 Academy Awards for his films? That's impressive! When his films debuted in the 1950s, it was the first time that most people had seen what goes on in the oceans. These days, most everybody has seen some sort of film with life beneath the ocean's surface. Many have gotten to swim in the ocean and experience it for themselves.

But in the 1950s, it was a whole new experience! Can you imagine getting to see giant sea turtles, octopus, dolphins, or even whales for the very first time? What an extraordinary encounter he was able to bring to everybody!

He did have a few crazy ideas, too. He predicted that sometime in the future humans might be able to have a surgery done where they could be given gills and live underwater. What do you think? Would you like to live like a fish? With all of the scientific advancements, maybe it could happen one day.

Jacque Cousteau was an interesting guy with a love for the ocean and a passion for sharing it. He was one of the first conservationists who spoke up for protecting the seas and all the creatures living in it. His love of the ocean was infectious and he was able to open people's eyes to all of the life deep beneath the water's surface.

Math Can Taste REALLY Good

What do math and dessert have in common? Well, that answer is as easy as pi! Yes, pi (the math version) or pie (the dessert version) both highlight the importance of circles and share the same delicious name. And along the way, some smart folks decided to celebrate the day with, you guessed it, all things pi.

So what exactly is the math version of pi? Pi or "π" as it is seen in the math world is a constant number that represents the ratio of a circumference of a circle to its diameter. That number is 3.1415 and keeps going on for infinity! People usually just use 3.14 as pi.

So what does this even mean? The circumference of a circle is how big it is all the way around. The diameter of a circle is how big it is across. A long, long time ago a famous mathematician, Archimedes, discovered that if you took the diameter of a circle and multiplied it by 3.14 it would equal the circumference. The cool thing is that this works, whether the circle is tiny or absolutely huge!

So you can see that this would be a handy tool to have for builders, engineers, and even space exploration. This type of math is handy even if you aren't a mathematician. But how does it relate to the dessert pie?

In 1988, the Exploratorium, a museum in San Francisco, California, decided to host a special day for their staff. Physicist, Larry Shaw, decided to host the event on March 14 and call it Pi day. If you write the date out in numbers it is 3-14, just like "π" is 3.14. Larry served pie and tea to the staff. Everyone loved it so much that they made it an annual event.

Larry's daughter realized another fun coincidence a few years later. March 14th is also Einstein's birthday. You know Einstein is a famous mathematician, so that makes the day even more fun.

It didn't take long for the news of a holiday that celebrates math with yummy desserts to spread and on March 12, 2009, the U.S. Congress declared it a national holiday. And while you may not get out of school for this holiday, there are lots of fun ways to celebrate. You could have a pie baking challenge, hold a pie fundraiser bake sale, and most definitely eat as much pie as possible.

Many businesses give a Pi Day discount. Bakeries and even places that serve "pizza pie" may run specials and be in on the fun. So what is your favorite pie? Is it apple? Or cherry? Or perhaps you really appreciate the math version "π" after learning all about it!

First You, Then Me

Marie Curie, one of the most revered women in scientific history, wanted to get a higher education so badly that she had to endure years of working as a tutor and a governess (nanny) in order to even begin to think about being able to afford a university education. Marie was not born into a wealthy family but her family did value education and learning, so they supported and shared her dreams.

Marie Sklodowska was born in 1867 in Poland, which was then part of the Russian Empire. Her father was a math and physics teacher and her mother was headmistress of a boarding school. Both of her parents were patriotic Polish, fighting for freedom from Russia which instilled a lot of Polish pride in their children.

The Russians would not allow women to attend universities. So when it came time for Marie and her older sister, Bronislawa, to graduate lower school, they had to think outside the box. For a while they attended Flying University, which was a secret school that met in changing locations around the city to provide Polish education for those desiring it.

Finally the two sisters struck up a pact. Bronislawa, the oldest sister, would head to a European university that allowed women. Marie would stay home and work to support and pay for her sister's education. After Bronislawa had graduated and established herself, it would be Marie's turn to attend university and her sister would financially support her.

The plan worked! Marie started out by offering tutoring, but soon realized she could make a lot more money as a governess. She worked for several years as a nanny to the young children of a sugar beet farmer. After 7 years, it was her turn to chase her dreams.

At the age of 24, Marie attended the University of Paris Sorbonne. In the beginning, she lived with her sister but decided to move so she could live closer to the university. She lived frugally, often sleeping in all of her clothes to ward off the cold and fainting from hunger when she would be too engrossed in her studies to eat. She started off the year behind in her studies, but worked diligently to catch up and finished first in her master's degree physics class her first year.

Searching for lab space to conduct research would lead to a life altering encounter. She shared lab space with Pierre Curie and the two fell in love and soon decided to get married.

Working together and studying minerals, the Curies made monumental discoveries for science. The first was the element polonium, named after Marie's homeland of Poland. This was followed by radium. Their work in radioactivity was awarded a Nobel Prize in physics in 1903, making Marie the first woman to win a Nobel Prize.

An unfortunate carriage accident killed Pierre at a young age, but Marie still continued with their work. She became the first female professor at the University of Paris and in 1911, was awarded her second Nobel Prize, this time in chemistry. This honor made her the first and only person to ever receive two Nobel Prizes in two different sciences.

Marie Curie's work changed and shaped the scientific community drastically, but it all started with a girl and her sister who looked out for each other and helped each other realize their dreams.

Someone Just Got an Earful

A man in eastern China went to the doctor and complained about a tickling sensation in his right ear. The doctor took out his endoscope (that's a tool doctors use to look in ears) and took a look. He was shocked to find that a tiny spider had taken up residence in this man's ear! Not only that, but the spider had created a web that took up nearly the whole ear canal. That gives me the creeps and makes my ears itch all at the same time.

The doctor was able to use the endoscope to get actual video footage of the spider/ ear invader. The spider was camped out near the tympanic membrane, or ear drum. The doctor tried to use tweezers to pull the spider out, but the spider was able to avoid being caught. Luckily, the doctor was able to use saline solution and squirt it up into the ear canal and flush the spider out. No damage was done to the ear canal, but I'll bet the patient had nightmares for a while.

In another instance of bugs in ears, Katie Holley had just moved to Florida and found it interesting that there were so many palmetto bugs everywhere. Palmetto bugs are another name for cockroaches. But in places like South Carolina and Florida, they are often referred to as palmetto bugs.

Katie woke up with a start one night with something strange going on in her ear. She went to the bathroom and put a Q-tip in there to fish around. When she pulled it out, she saw them. LEGS! Cockroach legs. Her husband did his best to get it out with tweezers, but they ended up going to the hospital emergency room for help.

The doctor dropped a numbing agent into Katie's ear and an anesthetic so there wouldn't be an infection. This killed the bug, too. The doctor pulled it out and sent Katie home with a crazy story to tell her friends.

Unfortunately, Katie's story wasn't over. Her ear still felt uncomfortable. A whole week later she was back at her doctor and they took a closer look thinking there was a bunch of ear wax that needed to come out. He flushed out the ear and it wasn't ear wax. It was SIX more pieces of cockroach! Even after all of that was out the doctor still wasn't convinced he had gotten it all, so he sent Katie to an ear specialist.

Later that day, Katie was with the new doctor. Surely there wasn't much left. But then out came even more legs, the head, and entire upper torso of the pesky critter. Katie's nine day ordeal was finally over.

Luckily, there are very few instances where other living creatures decide to get that up close and personal with your space. It seems that when this happens, it's a case of the bug looking for a warm place to sleep before looking for food again. This really isn't something for you to worry about. But it still probably makes you want to put in earplugs before you go to sleep!

Queens, Roller Coasters, & Coney Island

Have you ever been on a roller coaster? That feeling you get in the pit of your stomach as the coaster chugs up the hill right before you go soaring down and all around is pretty crazy, right? Roller coasters use inertia, gravity, and a lot of careful engineering to take people on the joy rides that they love.

The idea of the roller coaster actually came from Russia. With their freezing cold climate, the Russians in the 1400s built huge ramps. Some were 70 feet high and 100 feet long. These wooden slides were then covered in ice. The thrill seekers would sit on an ice block and Wheeeee!!!!! Some people reached 50 miles an hour.

Catherine II of Russia loved this so much that she had one constructed on her property. This one was built for year round enjoyment with a little track and wheels that could work without the ice. Europeans loved this idea so much that they tinkered with it and by 1817, Paris had its very own roller coaster.

Fast forward to 1884 on Coney Island in Brooklyn, New York. LaMarcus Adna Thompson is fed up with society. He feels like everyone is sinful and unholy. He wants to come up with a way to inspire people to be good. So...duh, he builds a roller coaster! (Not what you were expecting, right?)

Japanese quail represented birds. A few fish, brown shrimp and oysters represented fish and shellfish, with cockroaches and houseflies standing in for the bugs. Even plants such as tomatoes, tobacco, cabbage, onions, and a fern were grown in soil that contained the lunar matter.

Some of the organisms were injected with the moon matter, while some were fed with the ground up moon rocks sprinkled into their food. The quarantining astronauts were particularly interested in how the mice were doing. They knew that if the mice were doing well, they would likely be released from their quarantine and returned to their normal lives on schedule.

Luckily, no organisms seemed to be any worse for the wear after their encounter with the moon rocks. A few of the oysters died, but the scientists chalked that up to the fact that the experiment had been run during mating season and not any ill effects from the moon matter. Some of the plants even thrived more than normal when they were grown in the moon soil.

NASA's belief that the lunar matter was not harmful seemed to be holding up. Armstrong and Aldrin were released from quarantine. NASA scientists still performed this research for the next two space missions, Apollo 12 and 14. But finally, after Apollo 14, they were able to conclude that the moon was safe and that they could stop conducting these tests. I wonder if the cockroaches miss their moon meals!

The Surprising, Secret Life of Isaac Newton

Have you ever heard of the famous mathematician and scientist, Sir Isaac Newton? He's the guy who first understood and talked about gravity. Isaac is one of the most influential scientists of all time. He was brilliantly smart, an actual genius, but he was still just a human being like you and me.

Isaac was born in England on Christmas Day, and he was born really early, meaning he was very tiny. His mother claimed he would fit into a quart sized mug! It was surprising that he survived, as back in the 1600s they lacked the medical equipment necessary to help premature babies survive.

But Isaac survived and thrived as a child. He had a bit of a wild streak and even got expelled from school. Can you imagine this genius being kicked out of school?! He was twelve years old and got in trouble for fighting.

His mother thought he should be a farmer, but he hated it so when he was 19 he went away to college at the University of Cambridge. He discovered how much he loved mathematics and by 27 years old he was even a professor. He was asked to join Parliament (the English government) as a representative of Cambridge. However, he wasn't very successful as a politician. He was only recognized around the building for always complaining and being grouchy about the cold and the only time he ever really spoke up during a meeting was to ask somebody to close a window.

After he left Cambridge, he got to work doing his own experiments and making many important discoveries. It is said that he observed an apple falling from a tree and realized that "what is up, must come down" and his theories about gravity were formed. He also did a lot of studying on force and energy. Perhaps you've heard, "An object in motion stays in motion and an object at rest stays at rest." All bits of wisdom from our friend, Isaac Newton.

Later in his life, he was named the Master of the Royal Mint, which dealt with the money in England. Typically, the Master of the Royal Mint was more of a ceremonial job. Most previous masters hadn't been involved in the day to day work. But not Isaac! He was greatly bothered by people making counterfeit money. Counterfeit money is fake money that looks real. Isaac set to work to stop the people making it. He hired spies and did a lot of spying himself. He would dress in disguise and visit the seedy areas of town gathering information. It took several years, but Isaac was able to help catch and stop many of the best counterfeit coin makers of the day.

Isaac was knighted as Sir Isaac Newton by Queen Anne in 1705. He lived until he was 84, when he died peacefully in his sleep. People still marvel at how smart he was and how many great scientific and mathematical discoveries he was responsible for finding, but it's also really fun to learn all the other things that interested him.

Can Dead People Talk if They Have Golden Tongues?

In late 2021, archaeologists made an important discovery. They uncovered the tombs and remains of three ancient Egyptians. Although it's cool anytime ancient tombs or remains are found, this discovery is especially remarkable because one of the tombs, the tomb of the man, was untouched. This meant it had never been disturbed by grave robbers and that it was likely as intact as it was nearly 2500 years ago.

Esther Pons Mellado is a co-director at the archaeological mission there and according to her, "This is very important, because it's rare to find a tomb that is totally sealed."

The tombs were discovered at an archaeological site known as Oxyrhynchus, which is 100 miles south of Cairo, Egypt. This is in the location of an ancient capital city. So this means it is now an important area for archaeological discoveries.

The man's tomb contained a mummy in a casket with a man-shaped lid, 4 jars designed to hold the body's organs, and 400 funerary figurines made from faience, or glazed ceramic. The tomb of the woman and a three year old child were also found, but unfortunately had been previously disturbed by grave robbers. The robbers had taken anything of value and caused decay to those two mummies.

Perhaps the most extraordinary thing about these discoveries is that each of the three mummies were outfitted with a gold tongue. Now why would they want or need a golden tongue? It is all scientific speculation, but researchers think that embalmers taking care of the bodies of the dead would create these golden tongues so that the dead would be able to speak when they made it to the afterlife. They might need to be able to speak to Osirus, the god of the underworld. The research team is unsure if these people had speech impediments or why the tongues needed to be made out of gold.

Another mummy with a golden tongue was found earlier in 2021 near Alexandria, Egypt. Researchers say that mummies with golden tongues have only been found in Alexandria and Oxyrhynchus.

Archaeologists are uncertain who these mummies were, but they are hoping that as they excavate and find more, they will be able to piece together the mystery of the three bodies and their golden tongues.

Lucky Peanuts

Back in the 1960s, NASA was working hard on their space exploration program. Before they could even think about getting a man on the moon, they had to launch a spacecraft that could simply reach the moon. In 1961, they launched the first in a series of satellite probes called the Ranger Project.

NASA's Jet Propulsion Laboratory (JPL) was heading up these missions. The point was to get one of the Rangers to make it to the moon so it could take pictures. The mission turned out to be a tough one. The team launched 6 unsuccessful Ranger missions. It was finally on Ranger 7 that things turned around for the crew and they had a successful moon landing. Was it attention to details of past mistakes and careful engineering that helped this happen? Maybe. Or maybe it was lucky peanuts.

On July 28th, 1964, Dick Wallace, the space mission's trajectory engineer, passed out bags of peanuts to each of the crew members in the mission control room. "I thought passing out peanuts might take some of the edge off the anxiety in the mission operations room," he said. "The rest is history."

Ranger 7 had a successful mission and made it to the moon. And the JPL had a new and delicious tradition, lucky peanuts. Peanuts have been present for nearly every critical mission control event since this launch. They are even part of the unofficial countdown checklists. They are very important!

Normally it's bags or jars of regular peanuts that accompany the missions, but during a launch on Christmas Eve 2004, the peanuts came in the form of red and green holiday peanut M&M's. Luckily, these worked as well.

While the NASA engineers aren't superstitious about the peanuts, it's a fun tradition that unites years of people working together on space exploration. And who doesn't love a salty snack when they are working hard?

Meet the Giant Snake That Ate Dinosaurs

Scientists call it...the Titanboa. If the name, Titanboa, sounds scary to you, you're very smart. It WAS scary! These things would grow to 50 feet long and weigh over 2,500 pounds. That's as long as the school buses that go to your school every day and almost as heavy as your parent's car. There's no other way to put this. The Titanboa...was a monster.

But you don't have to worry about this giant snake. It only lived millions of years ago. And we haven't known about them for very long. Their fossils were just discovered 15 years ago. There is a huge coal mine in Columbia which is a country in South America. It is one of the biggest fossil discovery sites on the planet. But scientists may have completely missed the fossils if not for one college student.

Fabiany Herrera was studying geology at college and went to the coal mine with his class on a field trip. Fabiany was looking at some rocks on the ground and noticed that they all had impressions of ancient, prehistoric leaves on them. He showed the rocks to a scientist who worked there and the scientist got really excited. He called the Smithsonian and they sent paleontologists to Columbia right away. This student's discovery led to the great fossil hunt at the coal mine.

No land animal fossils had ever been found in this part of South America before. And once the scientists started looking, there were SO MANY! They found huge, giant turtle shells that weighed 300 pounds and 16 foot crocodile fossils with massive heads and jaws. And lots of bones were labeled as belonging to crocodiles. Millions of years earlier, this area had been a really big swamp. But some fossils that were labeled as crocodile, actually belonged to a giant snake. Scientists had no idea that there were ever snakes that big, so they didn't realize what they had found. It was only years later that they realized these massive bones belonged to the largest snake the world had ever seen.

The paleontologist, Johnathan Bloch, describes his mistake in this way. "My only excuse for not recognizing them is that I've picked up snake vertebrae before. And I said, "These can't be snake vertebrae." It's like somebody handed me a mouse skull the size of a rhinoceros and told me "That's a mouse." It's just not possible."

Fossils from 28 different Titanboa have been discovered since the first vertebrae was correctly identified. And every single one of them were over 42 feet long. This was obviously not a rainforest that you would have wanted to live in. For reference, the biggest snake in the world, the giant Anaconda, usually doesn't get any longer than 15 feet. That's plenty big for my taste but the Titanboa was in a completely different league.

Usually, it is very unfortunate when an animal goes extinct and vanishes from the world. But in this case... it's probably okay that we don't have to worry about 50 foot snakes that can swallow us in one gulp.

The Wright Brothers

Have you ever been in an airplane? That feeling of being up in the clouds is pretty amazing, isn't it? The Wright brothers sure thought so as they were the first to create and fly an engine powered airplane.

The brothers had an interest in aviation from a young age. Their father gave them a toy like a helicopter that used a rubber band to twirl its blades. It was their favorite and they became determined to make a similar life size machine that they could both use to fly.

They had several other jobs but they always worked well together. They owned a printing press, a bicycle repair shop, and even created custom bicycles. They were pretty creative guys! In 1899 they started working on their dream project...the airplane.

The brothers moved from Dayton, Ohio to Kitty Hawk, North Carolina. The ocean sand dunes at nearby Kill Devil Hills were perfect for flying, with gentle ocean breezes and miles of sand for a soft landing. In December of 1903, the Wright brothers thought they had a machine worth attempting a launch. The Wright Flyer had wooden propellers, a gas engine, and fabric wings.

So how did they decide which brother would be the first to go up in the aircraft? A coin toss, of course. Wilbur won the toss, but the first launch was unsuccessful. They took the plane back to their shop where they made the necessary repairs. Three days later, on December 17, 1903, Orville lay at the controls with Wilbur running alongside for lift off. The aircraft lifted into the air and glided for 120 spectacular feet. All of their hard work was paying off! Each brother took several turns piloting the aircraft that day. Surprisingly, that was the only day the Wright Flyer was flown. At the end of the day, a huge gust of wind caught one of the wings of the aircraft and it flipped several times. The damage was so great that the Wright brothers deemed it unrepairable. But it didn't matter. They now had an understanding of what was necessary to make an aircraft fly.

The Wright brothers went on to study more about aviation and tinker with their design. Surprisingly, they only flew together in a plane one time. They had promised their father, who feared losing both sons at the same time, that they would take turns in their aircrafts. They only went up together one time, a 6 minute flight in Ohio in 1910. Upon landing, Orville took their father for his first airplane ride.

The Wright brothers are responsible for changing the way people thought about traveling. They really shaped the way for modern aviation. They changed the way we see the world...and beyond. In 1969, Neil Armstrong and Buzz Aldrin flew to the moon and landed on it. That was only 65 years after the Wright brothers took their first ever flight. It's incredible that in just one lifetime, we advanced from the Wright Flyer to a space shuttle!

Speaking of Neil Armstrong, when he did his moon landing in 1969, he took with him a piece of the Wright Flyer for luck. In his pocket he carried a bit of fabric from the wing of the Flyer and a small piece of wood from the propeller.

So the next time you go up in an airplane, or even just see one in the sky...think about those two brothers working together on a deserted beach, just trying to make that plane glide for 120 feet. What an incredible journey aviation is!

NASA's Unknown Math Whizzes

Can you imagine being so good at math that you could take the computer equations designed to launch a man into space and do them by hand at your desk? That's exactly what Dorothy Vaughan, Katherine Johnson, and Mary Jackson could do.

These female mathematicians were sometimes known as human computers, or people who could perform complex math equations and calculations using just their brain, a pencil, and a slide ruler. They worked for the NACA, or the National Advisory Committee for Aeronautics, which later became known as NASA. They worked in the Langley laboratory in the West Area Computers section.

What was West Area Computers? During the 1950s, when these women were hired, the United States still had laws of segregation which meant that Black people and White people worked in separate spaces. The West Area was where the Black female human computers worked. Fortunately, when NACA transitioned to NASA, they did away with separate work spaces based on race.

Let's dig into each individual woman and learn a little more.

Dorothy Vaughn was the first Black NACA supervisor. She worked hard to make sure that her Black, female employees received fair wages and pay raises. When the space program began to use electronic computers, Dorothy knew this was the way of the future. She figured out Fortran, the computer's programming language and became an expert.

Katherine Johnson could make the precise calculations that were mandatory to make sure a spacecraft got to where it was supposed to go. She figured out the precise trajectory to help Alan B. Shepherd become the first American in space. She was on the team that performed the necessary calculations to put Neil Armstrong and Buzz Aldrin on the moon. Famous astronaut, John Glenn, asked for her to re-check the calculations for his orbit around the earth. By this point, the calculations were performed by computer, but Glenn felt better knowing Katherine had double checked them.

Mary Jackson worked at NACA and later developed an interest in engineering. Since it was during the time of segregation, she had to obtain special permission to attend classes with the White students. She persevered, received her engineering degree and was hired by NASA as the first Black female engineer. She later became the manager of the women's program at NASA, working hard to grow opportunities for women in the space and engineering program at NASA.

It's inspirational to celebrate these women, who were not only incredibly skilled mathematicians, but were also brave and brilliant leaders who demonstrated the power of equal rights for everyone.

Mystery in the Mountains

Hidden away in the Andes Mountains in Peru, sits an archeological and engineering wonder...Machu Picchu.

Machu Picchu was built by the Incas in the mid 14th century and abandoned about 100 years later. It sat abandoned for hundreds of years, only known by several local farmers when it was "re-discovered" by Yale archeologist, Hiram Bingham, in 1911. Hiram was searching for another abandoned city of the Incas when he came across Machu Picchu.

The Incas who built Machu Picchu were incredible engineers for their time. They were also keen observers of astronomy. Many people think that Machu Picchu was built as an astronomical observatory. One of the many interesting things you can see if you visit Machu Picchu is the Temple of the Sun, a giant rock, sculpted and believed to have been used as a clock or a calendar.

The other fascinating thing about Machu Picchu is how it was built. It was literally built into the hillside of the mountain. It sits 7,970 feet above sea level and covers 80,000 acres.

The Incas chiseled much of the granite directly out of the mountains around them and used it for the construction. A combination of palaces, temples, tiered gardens, and other necessary infrastructure were all constructed without the use of wheels or any steel or iron tools to assist in mining. That's a lot of heavy lifting!

The whole thing was built without the use of any mortar or cement. Each rock was carefully chiseled at a precise angle and wedged together tightly. This gave the buildings a definite engineering advantage. The surrounding areas have frequent earthquakes and Machu Picchu itself sits on top of two fault lines. Without the use of mortar (which makes things stick together) the stones can bounce around in an earthquake and then fall back into place. Researchers think that this is the reason that Machu Picchu is so well preserved today.

It's a great mystery as to why the Incas abandoned Machu Picchu. Some think it was lack of water, or the influence of Spanish conquistadors and their diseases. Whatever the reason, they left behind an abundance of interesting things that archaeologists and engineers are still studying today.

The Night the Lights Went Out in Times Square

The sun had set on November 9th, 1965. People were on their way home from work, cooking dinner, watching the evening news when all of a sudden the lights began to flicker. Then...complete darkness.

The power outage started in northern New York State near Canada, but quickly spread into parts of Connecticut, Massachusetts, New Hampshire, New Jersey, Pennsylvania, Vermont, and Rhode Island. Within 10 minutes, one of the most lit up places in the world, Times Square in New York City, was shrouded in darkness.

Below the ground, 800,000 people who were riding the subways home from work were trapped and forced to use flashlights to navigate the dark subway tunnels on foot. Remember, this was before cell phones. Today, nearly everyone would have a flashlight in their pocket on their smartphone. In 1965, most people weren't carrying flashlights around.

Elevators in the giant skyscrapers stopped working leaving people stuck inside way up above the city. Doctors in hospitals were forced to continue operating with only flashlights or battery powered lanterns as their only source of light.

The cause of all of this chaos? An overloaded power grid and human error. There was a safety relay in a power system near Niagara Falls that someone had set too low. When people got home from work and started turning on lights and heat, a power surge caused this safety relay to trip. This means that the flow of electricity was stopped in order to prevent overheating and other problems. This transferred the power to other electrical grids, but then they too, became overloaded and they tripped. The result was one after another power grids became overloaded and shut down.

Electric companies immediately got to work. Power grids that had not gone out were used to help restart all of the affected grids. And just like the domino effect that had caused the outage, power slowly started to be restored to all of the areas that were in darkness. By 7am the next morning, over 13 hours later, all the power was restored.

Over 30 million people and 80,000 square miles were affected by this power outage. Surprisingly, it was one of the lowest nights of crime in New York City's history.

The Worst Casting Job in Hollywood History Actually Killed Most of the Movie's Actors

John Wayne was one of the most legendary actors of his time. Usually best known for his role as the ultimate cowboy hero, he starred in over 170 films, even winning an Academy Award for his role in True Grit.

But in 1956, someone thought it would be a good idea for John Wayne to play the role of the legendary Mongolian warlord, Ghengis Khan in the movie, The Conqueror. But John Wayne wasn't Mongolian so the whole thing was really embarrassing. The movie was simply terrible. There's no question about it. But was it also deadly? Of the 220 people who worked on the film, 91 got some form of cancer. That's about 41 percent.

So what did filming a movie have to do with getting cancer? Plenty of people had made awful movies before and stayed healthy. The culprit was the location where the director chose to film the movie. The director thought that Utah would be the perfect setting for a movie set in Mongolia. St. George, Utah was a small town that was close to Snow Canyon. The director thought this place would be perfect for many of the scenes that they filmed throughout the movie.

The problem was that 100 miles away in Nevada, there was an atomic bomb testing site. In 1953, there were eleven above-ground nuclear tests that took place. The government had cleared the town of St. George and told the film crew that the area was safe. But unknown to everyone, this area that was downwind of the atomic bombing zone had become a radioactive hot spot.

The crew spent months onsite filming the movie. Later, the director had over 60 tons of dirt excavated from the site and shipped back to Hollywood to help make some of the re-shoots look more realistic. All of this exposure to the contaminated air and soil would have lasting effects.

The director of the movie died of cancer only 7 years after the release of the movie. In total, 91 of the 220 member crew got some form of cancer and 46 people died. If you worked on this movie, you had a 1 in 5 chance of dying from cancer. That's pretty high odds.

A doctor studying radiological health said that it seemed logical that the deaths were tied to the filming location and not just coincidence. He said that typically in a group of 220, you would expect to see around 30 people get cancer. The fact that 91 people suffered from cancer led him to think it was related to the environment.

From the time it was a nuclear test site in 1951 up until 1992, the Nevada test site saw 928 nuclear explosions. It is nicknamed the most bombed place on earth. Although they are no longer detonating any nuclear bombs there, it is still an area for weapons testing and the storage of dangerous waste materials.

John Wayne ended up suffering from lung cancer and later died from stomach cancer. To be fair, the lung cancer probably came from his habit of smoking 120 cigarettes every single day. Talk about your all time bad ideas. But the stomach cancer was a likely result of the radiation that he was exposed to from that terrible movie.

The Brooklyn Bridge & a Celebratory Rooster

The Brooklyn Bridge is a spectacular piece of the New York City landscape and was an incredible feat of engineering, especially considering construction on it started 150 years ago. The bridge spans 1595 feet and connects Brooklyn to Manhattan across the East River. It was the very first steel wire suspension bridge to ever be built and at the time was the longest suspension bridge in the world.

John Roebling was an engineer who had big ideas. He had already built several other suspension bridges when he came up with the idea to use steel cables to support the weight. Everyone thought the span between Manhattan and Brooklyn was too long to connect with a bridge, but John believed it was possible.

John convinced his son Washington Roebling, who was also an engineer, that this idea could work. They took their plan to The New York and Brooklyn Bridge Company and convinced them, as well. With the funds in place, construction on the bridge started in January of 1870.

Unfortunately, while conducting some measuring, John crushed his foot and had to have part of it removed. A few weeks later, he got tetanus and died. Washington took the job as chief engineer of the project.

The bridge was built using caissons on both ends. Caissons are large cylinders that use air pressure to keep the water out so that construction can take place underneath the water. However, the underwater air pressure was so extreme that it caused the workers to become ill. In less than six months, a project doctor treated 110 cases of this illness and named it Caisson's Disease. Three workers even died.

Chief engineer, Washington, suffered from a bout of Caisson's disease that left him partially paralyzed. Unable to return to the bridge, he viewed the construction through his telescope, perfecting the plans at home and relaying instructions to the crew. His wife, Emily Warren Roebling, was very smart in both the mathematical calculations for the construction as well as bridge specifications. Emily not only was the liaison between her husband and the crew members working on the bridge, she essentially spent the next 11 years acting as chief engineer and project manager.

In May of 1883, a grand celebration was held for the opening of the bridge. Along with many speeches, special performances, and an hour-long firework display, Emily rode in the first carriage to cross the bridge, carrying a rooster in her lap as a sign of victory.

The Brooklyn Bridge changed the landscape of New York, not only in looks but allowing a much easier way to travel. And it was all thanks to the three Roeblings, who didn't give up on the project, and all of the many workers who labored for so many years. If you get a chance to go see the bridge in person, do it! It's an incredible sight.

Sleeping in Space

Do you think that you might want to be an astronaut? One of the tests performed to see if you are cut out to be an astronaut has to do with testing for claustrophobia, which is the fear of confined spaces. During his NASA training, astronaut Scott Kelly said he had to curl up in a tight ball inside a rubber bag with a heavy zipper. They put the bag in a closet and left him with no idea of how long he would be in there. Scott wore a heart monitor to see how he handled the stress.

So why do astronauts need to be able to handle tight spaces? Because when you are up in space, you are lacking in, well...space! Whether you are in a space shuttle or in a space station, you will be confined to tight quarters both for sleeping and for everyday life.

If you are sleeping in a space station, each astronaut has an assigned cubicle that is about the size of an old phone booth. But if you are on a space shuttle, there is even less room. The astronauts all sleep in one area together.

So how do you keep from floating around? In space there is no gravity, so there is nothing to pull astronauts to their bed. They have to sleep with their sleeping bags tied to the wall to keep from floating away. Scott said that he missed having a pillow as it would just float away from his head. So he figured out a way to velcro himself to a cushion so it felt like he had a pillow.

Most astronauts sleep with earplugs and eye masks. It can be really loud in a spacecraft and because they orbit at such high speeds, astronauts will often see 15-16 sunrises per day. That can definitely be distracting when you are trying to sleep.

But not all astronauts have trouble with the noise and light. Astronaut Nicole Stott spent more than 100 days in space. "Sleeping in space was absolutely the best sleep I've ever had in my entire life."

"I always slept on the ceiling because where else can you sleep on your ceiling? You float into that bag and you find your position, and I would not wake up until the alarm went off."

So do you think you have what it takes to sleep in space? At least you have a few things to practice now.

Look Out Below!

Have you heard that if a penny were dropped off the Empire State Building, it would kill the person that it hit on the ground below? That is a tale that has been going around for awhile now and certainly makes you stop and wonder, and perhaps walk a little faster if you are passing under a skyscraper.

The thought is that as gravity pulled the penny back toward the ground it would pick up so much speed that it would become deadly as it crashed to earth. The Empire State building is 1250 feet tall. That's almost a quarter of a mile!

A physics professor from the University of Virginia, Louis Bloomfield, hosted an experiment that would test this theory. He held the experiment in a park, where he launched a bunch of pennies in a balloon filled with helium. He let the balloon soar hundreds of feet into the air and then used parts of a remote control airplane to drop the pennies to the ground.

It turns out that even though gravity is acting on the penny, air resistance is also acting. Meaning the penny, which is not very aerodynamic, is slowing itself down as it travels. This is created by drag which acts opposite of gravity. The penny will actually travel at a constant speed the whole way down.

So, Professor Bloomfield found that being pelted with these pennies wasn't a big deal. "The pennies didn't hurt," he said. "They bounced off me and it felt like getting hit by bugs, big raindrops, or little hail pellets. No bruises, no injuries. I was laughing the whole time."

So while there might be a lot of things to worry about while walking around a big city, it turns out that falling pennies aren't one of them.

Memorial on Mars

I'm sure you've heard about September 11th, 2001. It was a tragic day where terrorists attacked the United States and killed thousands of innocent people. It was a very sad day in U.S. history and changed the lives of many people.

On the morning of September 11, 2001, Steven Gorevan was heading into his office in lower Manhattan. As the founder of Honeybee Robotics, he and his fellow engineers were working on a special project for NASA's Mars Exploration Rovers, named Spirit and Opportunity. They were building "rock abrasion tools" used to grind away at a rock's surface and reveal the matter below. Their project was to be used to help explore and learn about Mars.

Then, September 11th happened.

Although none of the Honeybee staff were injured, they struggled to return to some sense of normalcy. Steve Kondos, one of the NASA engineers working on the project, had a great idea. Could they somehow incorporate some of the metal left from the wreckage of the Twin Towers into their design as a way to memorialize all of the people lost on that tragic day?

Someone was able to get in touch with the New York City mayor's office and within a few days they received a box. 'Here is debris from Tower 1 and Tower 2' was on a note accompanying the two pieces of twisted aluminum.

The team used these two pieces to build two shields to protect the rock abrasion tools while drilling, an American Flag decorating each.

Spirit and Opportunity, the Mars Exploration Rovers were launched in 2003 and remained working on Mars many years past their projected timeline.

"It's gratifying knowing that a piece of the World Trade Center is up there on Mars. That shield on Mars, to me, contrasts the destructive nature of the attackers with the ingenuity and hopeful attitude of Americans," said Steven Goreman.

Although the Rovers are no longer operating, they remain on Mars as a reminder to the past and a nod to the future.

A Man Who Wasn't a Doctor, Blue Babies, and the Invention of Heart Surgery

This is the story of Vivien Thomas, a Black man who changed the course of medical history. In 1930, Black people could not become doctors. There was a lot that Black people were not allowed to do because of the color of their skin.

Vivien was 19 years old in 1930 and was working as a carpenter's assistant in Nashville, Tennessee with plans to go to college. He was also interested in a career in medicine. The Great Depression was in full swing in the early 30s which made life tough for everyone. Vivien had lost his savings and needed a job.

A friend told him about an opening at Vanderbilt University for a laboratory assistant. The doctor who needed the assistant was apparently very difficult for anyone to get along with. That young doctor was Alfred Blalock. Alfred and Vivien would work together for the next 30 years and together, they would have an enormous impact on the history of medicine.

They quickly developed a mutual respect for one another when Vivien was hired for the job. But because Alfred was White and Vivien was Black, they could not be friends outside of the laboratory. Inside the laboratory, the two men were accomplishing amazing things.

Alfred had been studying traumatic shock when he hired Vivien, which is what happens to the human body when it is badly injured. Working with Vivien, Alfred's theory on shock became the foundation for how doctors understand the effects of traumatic shock.

Vivien Thomas became one of the most important medical researchers in the world in just 4 years with Alfred. And he didn't even have a college degree! But because he was Black, he was classified as a janitor by Vanderbilt University and was not paid the same as White researchers and lab assistants.

Vivien spoke with Alfred about this and got a raise but he was probably still classified as a janitor according to those who sent out the checks. Alfred would get job offers from other hospitals and told them that he wouldn't go without his partner, Vivien Thomas. Those hospitals then took back their offers because they were not willing to hire a Black man to do anything other than to pick up trash.

In 1941, the early days of World War II, the two men would move to Johns Hopkins which was an extremely reputable research hospital. Alfred was the new surgeon in chief and Vivien moved his family from Nashville to Baltimore. They also took 5 dogs who had hearts that the two men had rebuilt. This was groundbreaking surgery for 1941.

Vivien Thomas' first day walking to work as a Black man in a surgical coat caused quite the commotion.

People had never before seen such a thing and Vivien had to be brave. But he did what he always did. He got right to work.

A big problem at this time was babies being born "blue". Their arteries and veins weren't right which didn't allow the blood to pump through the heart correctly. The two men became obsessed. Not many doctors had done surgery on the heart at this time. Vivien and Alfred were breaking new ground but were confident that their work with the dog hearts held the answers to this problem with blue babies.

In 1944, they tried the first surgery to save the life of a "blue" baby. It was Vivien who had been the main surgeon on the dog hearts. But it would have to be Alfred performing this procedure because Black men were not allowed to perform surgery. Vivien wasn't sure he wanted to participate but Alfred told him he needed him there. So Vivien stood on a stool so that he could see over Alfred's shoulder.

When operating on a baby, everything is smaller, the size of the heart, the arteries, everything. This makes surgery much more difficult. Alfred asked Vivien questions constantly as he worked. He knew that Vivien was the real expert here. Vivien calmly reassured him and answered his questions as he went. Alfred opened up the baby's chest and cut the pulmonary artery and the subclavian artery and switched them where they connected to the heart. This had never before been done on an adult much less a tiny newborn baby.

Sometimes Alfred would start to make a cut in the wrong direction and Vivien would correct him.

Then the moment of truth arrived. Alfred removed the clamps stopping the blood flow and right away, the little blue baby turned pink. It worked! Vivien described it as a miracle. Word spread quickly and blue babies from all over the country came to the Johns Hopkins miracle workers. It was nonstop. Surgery after surgery, little lives were saved everyday. And everyday visiting surgeons were baffled by the Black man (who wasn't a doctor) who stood behind Alfred telling him what to do.

While Alfred was a groundbreaking scientist, he was not a skilled surgeon. He would be the first to admit that fact. This huge advancement in medical science, the invention of heart surgery, would not have been possible if not for the skill and expertise of Vivien Thomas. He was a master surgeon who was not allowed to actually touch patients on the operating table. It was Vivien who took Alfred's ideas and discovered ways to make them a reality in the surgery room.

Vivien Thomas never went to college. He never got a medical degree. But his legacy is that he became one of the most influential figures in medical history by essentially being the inventor of heart surgery. He was eventually given an honorary doctorate from Johns Hopkins and a great movie was made about him if you would like to learn more. It's called, *Something the Lord Made.*

Medical Students Battle Animal Rights Activists

When someone says they are "being a guinea pig," it usually means they are the first to try something new. This saying dates back to the 18th century, when guinea pigs were used in laboratory experiments.

For many years throughout history, medical and scientific advances have been made at the expense of animals. Unable to defend themselves and readily available, all sorts of animals have been used as testing for studying side effects.

While we are grateful for so many of the medical and scientific advances that animals have afforded humans throughout centuries of research, in the early 1900s people started saying, "Enough!" Over the last several hundred years people's relationships with animals have changed. A long time ago, most people did not keep animals that didn't serve a purpose. They were used for food or work. It wasn't until later that people began to keep animals as pets. Even then, only very wealthy people who had plenty of time and money would keep a pet. But by the early 1900s more and more people in the middle class were enjoying domesticated animals as companions.

So when a medical professor at University College London was put on trial for inhumane and illegal experiments on a little brown dog, plenty of the public were outraged. Even though the medical professor was found not guilty, anti-vivisectionists (people that were against experimenting on live animals) decided to do something about it. They erected a statue of a little brown terrier, in honor of the little dog in one of the progressive areas of London. The statue stood as both a human and canine watering fountain and had a plaque detailing the crimes against the dog and calling out University College London (UCL) for its part in the inhumanity.

While many people appreciated the statue, the medical students at UCL did not. They claimed the dog hadn't been mistreated and that UCL wasn't the only university still testing on animals. One night, about a year after the statue had been erected, a couple of medical students from UCL decided to take matters into their own hands and attacked the statue with a crowbar and a sledge hammer. Although they were stopped by a nightwatchman, it started the beginning of what became known as the Brown Dog Riots. Medical students from UCL and other universities took to the streets, fighting and marching against the anti-vivisectionists, suffragettes, and other progressive groups.

This went on until 1910, when politicians, tired of the drama over the dog statue, had it removed under the cover of night. The statue was hidden away in a blacksmith's shed and later destroyed.

People are still fighting for animal rights and in 1985, a new statue was built in Battersea Park, London and remains there today. It bears the original, controversial inscription on its plaque and advocates for humane treatment of animals while still advancing science.

This 19 Year Old Just Spent 155 Days Flying Around the World

Working hard and relentlessly chasing her dreams helped Zara Rutherford accomplish two of her goals. On January 20, 2022, she successfully became the youngest female to fly by herself around the world, bringing attention to female aviators everywhere.

Zara was 19 years old when she took off from an airport in Belgium in August of 2021. Her journey took her plane over 41 countries and five continents. The 155 day journey covered 32,000 miles.

Zara grew up in a family of pilots and aviation has always been important to her. One of her main goals is to encourage girls to pursue activities and careers that include STEM and aviation. Right now, only 5 percent of airline pilots are female. And that's something Zara wants to change. She's hoping to bring awareness to female aviators and prove that anybody can achieve their goals if they set their mind to it.

"The youngest man to have flown solo around the world, Travis Ludlow, was 18 years old. The youngest woman, Shaesta Waiz, was 30 at the time of her flight. With my flight I hope to reduce this gap from 11 years to 11 months," Zara says on her website tracking her journey.

And she did indeed narrow this gap. Zara's partner in her journey was a Shark ultralight aircraft, a very small plane designed specifically for distance flying. These are some of the fastest ultralight aircrafts and can reach speeds of 186 miles per hour.

Her journey took about two months longer than expected, due to weather issues and wildfires. But she said that seeing the world was amazing and beautiful. And for her next adventure? Zara wants to see more of the earth, but she wants to do it from space. She would love to continue working in aviation by becoming an astronaut.

This 19 year old is now the youngest female to circumnavigate the globe and she encourages others to go after their dreams, just like she did.

A Year in Space? Are You Even My Brother Anymore?

What do you do when you want to study the effects that being in space has on the human body? You find a set of identical twin astronauts and ask them to be part of your mission, of course.

That's just what NASA did with the help of ten research groups from around the country and twin astronauts Scott and Mark Kelly. Coined as the Twins Study, researchers were able to compare data from Scott, who went up into space, against Mark, who stayed on earth, and have a better understanding of how space travel affects the human body.

This project was unique because Mark and Scott have the same genetic make-up, so there were less variables than studying two unrelated humans. Obviously, NASA wants to keep their astronauts safe while they are traveling in space so this mission provided data that will be used for decades to come. Researchers are particularly interested in the effects of long-term space travel as a flight to and from Mars would take a whopping 3 years.

Launched in 2015, astronaut Scott Kelly spent 11 months aboard the International Space Station. He was the test study up in space, with researchers taking note of any physiological, cognitive, and molecular changes to his body. His results were compared to his brother, Mark Kelly, who was the control study living back here on earth.

The results? NASA was very pleased to learn that the human body can adapt quite well to the harsh environment of space. It seems you could go live in space without too many crazy side effects. Some of the data gathered may even help scientists develop treatments for stress related health problems here on earth.

But here's something interesting. They detected a change to Scott's chromosomes and DNA. What? Isn't our DNA what makes us...us?

Researchers can take a sample of your DNA and tell you who you're related to. So were Scott and Mark still brothers? Yes. These were interesting changes that happened to Scott's chromosomes but these changes reverted back to normal after 6 months of being back on earth. The brothers joked that they were back to being identical twins at that point.

Being an astronaut is a pretty unusual job, and it's even more unusual for twin brothers to BOTH become astronauts. In fact, the Kellys were the first relatives to both become astronauts. It's even crazier that they're also identical twins. So it was definitely a unique opportunity that NASA had to collect data that will hopefully help keep future astronauts safe.

The Oldest Rocks on Earth

Australia definitely sounds like a cool place. Kangaroos, koalas, the Great Barrier Reef...but did you know that along with being the home to cool animals, Australia may very well be one of the oldest places on earth?

The Pilbara region, located in northwestern Australia began to form 3.6 billion years ago. The Aboriginal Australians, who dwell in this area, are thought to be the oldest continuously living culture. They have been in existence for over 50,000 years, possibly even before the last Glacial Period, when ice sheets covered much of the world.

The Pilbara is special, and scientists and archaeologists are only beginning to tap into all the mysteries, and perhaps answers...that lie within.

The area is enormous at nearly twice the size of Great Britain, yet sparsely populated with only 61,000 residents. Much of Australia's wealth comes from iron ore that is mined in the Pilbara. The vast, dry region is filled with mile after mile of ancient rock formations, mineral deposits, red cliffs, and gorges that act like a time capsule. Here, layers and layers of rock can be chipped away to discover more of the earth's history.

In 1980, a stromatolite that is 3.45 billion years old was unearthed in the Pilbara. Stromatolites are fossilized evidence of the world's oldest life forms.

These micro bacteria were the major life form on earth for over 2 billion years and are likely responsible for why the earth has oxygen. Stromatolite fossils can be found all over the Pilbara.

Here is something else that's interesting. In 2019, NASA scientists preparing for a journey to Mars came to the Pilbara to study. Many had never seen such ancient evidence of early life forms. To study and learn the details of the stromatolite fossils will hopefully help them prepare for what to look for as far as life forms on the planet Mars. Geologists say that the chemical makeup of the Pilbara rocks and the amount of iron found in them is very similar to what is likely on Mars.

Research in the beautiful Pilbara is only beginning and it's exciting to think of what scientists will discover and learn as they journey deeper into this desolate corner of the world.

7th Grader Makes a Big Discovery

This story proves that kids can make BIG discoveries, even if they are just fulfilling their school science fair obligations.

Elan Filler was in 7th grade when she was looking for a science fair project idea. Her dad is an infectious disease specialist at the University of California. He was chatting at a conference when he was put in touch with Deborah Springer, from Duke University.

Deborah had been studying Cryptococcus gattii, or C. gattii for several years. C. gattii is a fungus that has been causing people to get sick. People with compromised immune systems are especially susceptible to getting sick from it. People in California had been suffering from this fungus for 10-12 years, but researchers couldn't track down how they were coming into contact with it. Trees were suspected, but they needed to find out what specific kind of tree was responsible. It was a mystery!

This is where Elan played an important role. Her dad put her in contact with Deborah and they came up with a plan. Elan took samples of all kinds of trees all over her hometown of Los Angeles. She grew the samples of fungus in petri dishes. Her first batch contained mainly eucalyptus trees, as they grow the fungus in Australia. But none of those samples tested positive for C. gattii. It was back to the drawing board.

Her next round contained samples from many other tree varieties. This time she had three samples that tested positive for the fungus. The Canary Island pine, the New Zealand pohutukawa, and the American sweet gum all grew the fungus C. gattii. They also matched the test results from the patients that were infected with C. gattii 10-12 years ago, meaning this has been an ongoing problem.

Elan had her name published as an author on the study when the official results were published. Her assistance with this research will help scientists continue studying C. Gattii and how to help those that have been exposed and prevent the spread of the fungus. Not bad for being a 7th grader doing a science fair project!

House of Bones

Do you know anything about the Ice Age? It was probably pretty cold! Many days the temperature was below zero degrees Fahrenheit. There were giant wooly mammoths running around. What about people? What was their life like?

Scientists have been trying to piece together what life was like for early humans during the Ice Age and in recent years have made some incredible discoveries that are helping them. In 2014 in Kostyonki, Russia, paleontologists unearthed a large structure, like a big house. Guess what it was made of? Mammoth bones!

The large structure is about 40 feet across and is made up of the bones of over 60 mammoths. Mammoths are similar in size to elephants, so many of these bones are quite large. Building a structure this size took a lot of planning and thought. Imagine a giant igloo shaped structure made entirely of bone! Pretty cool, right?

Paleontologists have found other huts made from bones, but those huts were usually only 12-15 feet wide. This new dwelling was a lot bigger. It's also one of the oldest structures that has been found anywhere. They think this structure is over 25,000 years old!

Researchers have a lot of questions they are trying to answer. What was this hut used for? There is evidence of a charcoal fire pit within the structure as well as preserved food scraps and smaller piles of mammoth bones. Most researchers think it was used as a ceremonial place. Others think the huts were only used as food storage. No matter what they think, researchers all agree that finding this structure shows that prehistoric man was more sophisticated than they originally thought.

To build this type of structure took time, planning, and strength. And where did they get 60 mammoths? This is another question the researchers are trying to solve. They think perhaps the area was near a water source that didn't freeze. That would make it a good area for animals, as well as humans. However they got there, the mammoths were vital to the survival of prehistoric man, providing food and now we know, shelter as well.

This finding is still relatively new, so it will be interesting to find out what else scientists learn as they do more research. One thing is for sure, I'm pretty glad my house is made of bricks and not bones!

What Are the Chances?

Probability is a math term for how likely an event or outcome is to occur. Let's take a look at an easy one. Grab a coin. The coin has two sides, heads and tails, right? So if you toss the coin, there are only two possible outcomes for how the coin will land. But they can't both happen. Only one will actually land face up. So the odds are 1 out of 2 that you will predict heads or tails correctly. So you've got a 50 percent chance of getting it right. Not too shabby.

Some mathematicians have used probability to predict some crazy outcomes. Let's take a look!

Odds of being born with 11 fingers or toes is 1 in 500. So for every 500 people you know, it's likely that one of them has an extra finger or toe.

The American Bowling Congress (did you know that such a thing existed?) thinks that the odds of scoring a perfect bowling game, meaning getting a strike 10 times in a row are 1 in 460 for a professional player and 1 in 11,500 for a person playing for fun. So only 1 in every 11,500 games of casual players will be a perfect game. I feel like I could probably bowl 50,000 games and not have a perfect game.

If you live near the ocean and like to swim, your odds are 1 in 3.7 million that you will die from a shark attack. However if you don't live near the ocean your odds are even less at 1 in 7 million. So it looks like you are probably safe from sharks no matter what!

What about the mega jackpot lottery? Are you feeling lucky? Well, lottery odds are pretty low. You have a one in almost 14 million chance of winning. Yikes! I don't feel so lucky after all!

And since this is a science book, what about the odds of becoming an astronaut? When NASA took applications in 2017, there were 18,300 people that applied for the job. Of those 18,300 people only 12 made it into their training class. Great opportunity for those 12 amazing people!

But where did all of this fun with probabilities get its start? The answer lies in Italy during the Renaissance era with one particular mathematician who loved gambling. I'm talking about Girolamo Cardano.

Giralamo was a brilliant man who invented the combination lock and solved some problems with algebra that he was the first to discover. He was also one of the first mathematician's to deal with negative numbers. But the poor guy had a gambling problem. He loved playing dice games and often lost all his money despite earning quite a bit for being such a famous scholar.

He lived during the 1500s and during his lifetime was sometimes referred to as "the gambling scholar." Gambling did not help Giralamo in his life. It cost him a lot of money and relationships. But it did lead him to be amongst the first scholars in history to study what is now known as probability theory.

During the 1500s, probability theory was nonexistent and unexplored. In dice games, people thought that with two 6 sided dice, you should roll a 10 as often as you roll a 9. What do you think?

You could try it and see for yourself in 50 dice rolls how many 9s you roll versus how many 10s. Or I can just tell you. Giralamo Cardano was the first person to discover that the reason that 10 isn't rolled anywhere near as much as 9, is because of the fact that there are less rolling combinations that add up to make 10. To roll a 9 with two dice, you can roll a 5 and a 4, a 4 and a 5, a 6 and a 3, or a 3 and a 6. There are 4 combinations to roll a 9 with two 6 sided dice. There are only 3 possible combinations to roll a 10. These would be a 6 and a 4, a 4 and a 6, and a 5 and a 5. There are fewer possible combinations resulting in a lower probability that you will roll a 10 versus rolling a 9.

During Giralamo's lifetime 500 years ago, this was a groundbreaking discovery. You'd think that this knowledge would have made Giralamo a better gambler. And maybe it did. This is how one smart guy who enjoyed gambling too much became known as the father of probability.

"Willebrord Snellius, You Get Down from There This Instant!"

If you had happened to be traveling along Dutch country roads in the early 1600s, you may have come across an unusual sight. You would probably have been startled to see a man peering out at you from atop a church steeple.

This man actually climbed up lots of church steeples in the Netherlands. But he was no adrenaline junkie, he was the mathematician and mapmaker, Willebrord Snellius. (This might be one of the coolest names you've ever heard. Can you imagine if your mom had named you Willebrord?)

Willebrord is credited with discovering one of the primary laws of physics, the law of refraction. But what does that have to do with climbing church steeples?

There weren't many tall buildings in the Netherlands in the 17th century. I suppose there weren't many tall buildings anywhere at that time. This made mapmaking rather difficult in places like the Western Netherlands which doesn't have any mountains either. It isn't even that hilly!

Willebrord figured out that the best way to measure distance was not to count your steps going from one place to another, but to get as high as you could and measure the triangulation from one place to another using a large quadrant. It wasn't as easy as it might sound. It took a lot of mathematical calculation.

He managed not to kill himself up there on these church steeples holding on tightly while he measured distances with his quadrant. His efforts resulted in the most accurate map of the Netherlands that there had ever been.

It was so accurate that the Dutch military kept it all to themselves and wouldn't allow it to be published. They found it so valuable, they didn't want it falling into the wrong hands, such as the pesky French who invaded the Netherlands several times.

But Willebrord was eventually given full credit for such accurate mapmaking and his legacy became one of great admiration among the Dutch people for his accomplishment in cartography (map making). He also used his prowess of calculation to figure out the circumference of the Earth!

Mind you, this was around the year 1615. This sort of thing hadn't even been attempted or thought about much for over 1,500 years. Willebrord's calculations were so precise, that he managed to come within less than 1,000 miles of the earth's true circumference. Less than 1,000 miles! That's only 3.5% off the mark. Not too shabby, especially for a guy most people probably thought was a bit weird for climbing up all their church steeples.

Kids (and Science) Help Solve a 300-Year-Old Shipwreck Mystery

The wind was howling and the cold, crashing sea began to tear the ship to pieces. Sailors scrambled everywhere to save their ship and their lives. They screamed commands at each other but the wind and the massive waves were deafening. Then the unthinkable happened. Fire. A lantern must have fallen and broken. The men desperately tried to put the fire out while trying to keep from being thrown overboard. But the huge ship was doomed.

The wild Pacific coast of Oregon is known as something pretty scary to sailors. It's spookily called the Graveyard of the Pacific. That's because lots of shipwrecks happened there due to the rocky coast, the dangerous currents, and the frequent storms. Local kids loved searching for treasures from these ancient shipwrecks that would sometimes wash up on shore. Craig Andes was one of those kids growing up on the Oregon coast. He was on one of his missions inside a sea cave when he saw it. It was a piece of wood. But Craig thought that it looked old, *really* old.

So he kept it. Years passed. Craig grew up and heard about the effort to find the Santo Cristo de Burgos, a ship that had gone missing 300 years earlier. The Santo Cristo had been traveling from the Philippines to Mexico in 1693 when it got blown off course by a storm. It was a massive ship carrying beeswax, along with silk and porcelain from China. There were no bees in North America at that time. So American settlers thought it strange when Native Americans in Oregon had all this beeswax to trade. Where did that come from? For hundreds of years, it was a mystery.

It was believed that there was no way that pieces of the Santo Cristo would survive 300 years of crashing waves in a sea cave. But Craig's piece of wood was studied by scientists at the Maritime Archaeology Society anyway and...WHAT? Craig's wood was from Asia (a long way from Oregon) and over 350 years old! This discovery in the science lab happened in 2020 and two years later, Craig went with researchers to the sea cave where he had found the wood as a boy. They were able to pull out more pieces of the Santo Cristo along with some of the Chinese porcelain and beeswax that had been onboard.

The mystery of the ancient Beeswax Wreck had been solved thanks to interested scientists and a curious young treasure hunter kid. That's always a winning combination.

Man, It's Windy Today!

What is the fastest growing mode of producing electricity in the world right now? Wind! All across the globe, engineers and renewable energy experts are finding ways to harness this powerful commodity and put it to good use.

How long have people been using the wind to create power? Probably a lot longer than you'd think. About 4000 years ago people in China and Persia began using the wind to pump water and grind grain. And of course, let's not forget about sailboats that have been using the wind as power for even longer.

The first modern wind turbine (giant windmill) was built in Vermont in 1941. Palmer Putnam was an energy conservationist with ideas ahead of his time.

Back in these days, most windmills were the small "western" style variety that you'd picture out on the open plains. They could maybe create enough power to charge some batteries or pump a little water. They certainly weren't capable of the task that Palmer proposed, which was to use giant blades to harness the wind and turn it into one megawatt of power. The power created would be fed into the nearby electric grid and work together with the nearby hydroelectric plant to create sustainable power for the area.

This was heady stuff. And many of these ideas were much grander than anybody at the time thought possible. But Palmer found investors and workers and got the project up and running. This was right before World War II hit the United States and they were racing to complete the project ahead of the war.

And so the largest wind turbine in the world was created. This turbine had blades that were each 75 feet in length and it was ten times more powerful than any turbine that had ever existed. In September of 1941, the turbine blades began to turn...using the mountain wind to create power for surrounding towns.

The turbine went through hundreds of hours of rigorous testing and was by most means a success. An unlucky bearing break in 1943 caused a disruption in its progress. By now, the United States was at war and metal was being rationed, meaning the bearing wouldn't get fixed until 1945.

Shortly after the turbine was back functioning, disaster struck. One of those huge blades became detached from the windmill and went sailing off into the night. Look out below! It landed 750 feet from the windmill. Unfortunately, this was to be the end of Palmer's wind turbine, as Congress deemed it a failure and by that autumn, the blades had been removed and the turbine torn down.

Palmer Putnam may have had ideas that were just bigger than the time he was living in, but his wind turbine was by no means a failure. It has been studied as people dive further into research for wind energy. And wind energy is a growing resource. Some researchers predict that by the year 2050, one third of the world's energy may come from the wind.

Four Year Old Girl Discovers 220 Million Year Old Fossil

Walking on the beach, you can expect to find some cool things. Pretty shells, washed up bottles, and other sea treasures are the usual beachcomber finds. But while taking a walk on a beach in Wales with her dad and pet dog, four year old Lily Wilder discovered something way cooler. A dinosaur fossil!

Preserved in a rock that was about up to her shoulders, Lily saw the exact footprint and pointed it out to her dad. It was about 4 inches long. Her dad took a bunch of pictures and thought it was neat. Later, Lily's grandmother encouraged them to share their find with fossil experts to see what they had found.

It turns out they found something that has scientists really excited! The footprint is thought to be that of a dinosaur that lived 220 million years ago. It's impossible to figure out exactly what species of dinosaur may have left the print, but scientists are able to determine a few facts.

The dinosaur likely stood on two rear legs, almost 3 feet tall and 8 feet long. It was slender with a long tail and likely dined on other smaller creatures and insects. The details in the foot, such as where the joints and muscles were located, were very well preserved. This dinosaur footprint will hopefully help scientists learn more about the way two legged dinosaurs walked.

The fossil was taken to the National Museum Cardiff for more scientific research and for others to be able to see and appreciate it. Lily is listed on a plaque as the official finder and she and her class will be invited to the museum to see it.

Hidden Pictures

In 1868 in the small town of Santillana del Mar in northern Spain, a man was out hunting with his dog when the dog became trapped in some rocks. The man freed the dog, and when he did, he noticed that the area opened up to a hidden cave. He informed the land owner of his findings, but ten years would go by before the cave was properly explored.

Ten years after the dog incident, Marcelino Sanz de Sautuola, had taken an interest in fossils and studying prehistoric times. He decided to go fossil hunting with his 8 year old daughter. Searching along this Altamira cave area, the father was fossil hunting while the little girl explored the cave. Suddenly she exclaimed, "Look Papa! Oxen!" On the ceiling of the cave were what appeared to be prehistoric paintings.

Marcelino was fascinated with all things prehistoric and he had just attended a show in Paris filled with prehistoric objects. The cave drawings seemed so similar. He decided to contact a researcher in Madrid. The researcher and Marcelino studied the paintings and even presented their findings at an international conference in 1880. The result? Many experts thought the theory that this was prehistoric art was ridiculous. They even accused Marcelino of fraud!

It wasn't until many years later, sadly after Marcelino had passed away, that similar paintings were found in France. Finally, the Altamira cave paintings had some context. It was determined that they were from the Upper Paleolithic era, meaning they were likely 36,000 years old!

This story is proof that sometimes with science, people make great discoveries that are ahead of their time. It's important to keep pushing the boundaries and stay true to your beliefs. And also...sometimes kids and dogs are the best explorers.

The Dark Lady of DNA

What if you spent years researching, building ideas and evidence, and becoming one of the best at your job, only to have someone else get the credit? Pretty unfair, right? Yet, that is exactly what happened to Rosalind Franklin.

Rosalind was extremely intelligent as a young girl and knew at age 15 that she wanted to be a scientist. Her father discouraged this, believing that because she was a woman, it would be extra hard for her to work in the scientific field. It turns out he was somewhat right, but we are still glad Rosalind was brave enough to do it anyway.

Rosalind studied at Cambridge University and had a series of research jobs. She preferred to work independently and found a few times where her peers or directors didn't take her as seriously because she was a woman.

Rosalind took a job in Paris where she was introduced to Jacques Mering. Jacques taught her X-ray diffraction techniques and she became extremely skilled with this form of technology. So much so that she was offered a research scholarship at King's College in London, setting up and improving their X-ray crystallography unit.

Rosalind would be working alongside Maurice Wilkins, another scientist who was working with X-ray crystallography. However, when she arrived to start work, Maurice was out of town on vacation. When he returned, he assumed that because she was a woman, she was only hired to be his assistant and treated her as less intelligent. Of course, this was not the case and it made Rosalind furious! They were off to a rough start in their working relationship.

During this time there were several scientists, including Rosalind and Maurice who were researching and trying to understand DNA. DNA is the molecule that makes up humans and every other living organism. It's like a map with all of the instructions on who you are. Rosalind was able to use her superior X-ray skills to take two photographs of DNA. What she found in the X-rays supported her thoughts on the structure and make-up of DNA.

Two other scientists were also researching DNA. James Watson and Francis Crick were also hard at work to identify and understand DNA. They would occasionally share information and ideas with Maurice Wilkins. Rosalind preferred to work without them. On one occasion, Maurice shared the X-ray images that Rosalind had taken with James and Francis. The images were just what James needed to take his understanding of DNA to the next level. It was the missing piece that supported what he had been thinking.

James and Francis soon after published their research, thus they were given the credit for understanding how DNA worked and what it looked like. Rosalind left her job at the King's college and began working on other projects, including research on the polio virus.

Rosalind died from cancer in 1958. She was only 37 years old. In 1962, a Nobel Prize was awarded to James, Francis, and Maurice for all of their research and studies of DNA. Unfortunately, the Nobel Prize committee doesn't award Nobel Prizes for people that are deceased. So it is only in more recent years that the world is becoming aware of the very important role that Rosalind Franklin had in the scientific discovery of DNA. We are grateful for her perseverance and dedication to science, even during a time when it was really hard to be a female scientist. For a very long time, very few people knew of her contribution to this groundbreaking scientific discovery, so she is often referred to as the Dark Lady of DNA.

How Heavy IS That Hairball???

Lexington, Kentucky is home to the state's largest college, the University of Kentucky. But that's not the only college in Lexington. There's also the spookily named Transylvania University there as well.

And if you walk along through the rolling and leafy green campus of Transylvania University, passing buildings like Hazelrigg Hall and the Cowgill Center... you just might stumble across a building with a really peculiar oddity lurking within its walls. Let's enter Transylvania's Moosnick Medical & Science Museum.

It is here that you can cast your eyes upon...wait for it...THE LARGEST HAIRBALL IN THE KNOWN WORLD!

This is interesting for lots of reasons. But first, let's dig into how they got it. It was actually given to the university by Abraham Lincoln's brother-in-law. What the heck was he doing with such a weird curiosity?

Nobody really knows. But Lincoln's brother-in-law was George Rogers Clark Todd, the first lady Mary Todd's youngest brother. George graduated from Transylvania's medical school and gave the hairball to the school as a reverse graduation gift.

It was long thought to have come from the stomach of a buffalo. It's bigger than a bowling ball at 14 inches across so it sure didn't come from a cat! It actually came from the belly of a cow and it must have been really uncomfortable for the poor cow to carry around.

One interesting thing about hairballs that you may not know, is that they were once used as a medical treatment. Abraham Lincoln actually gave a small hairball to his son to swallow after he was bit by a rabid dog. Would you swallow a hairball?

People thought that they would soak up poison and they would pay a LOT of money for them. Fortunately, Lincoln's son survived the rabies bite. But it certainly wasn't because of the hairball. Other remedies for rabies bites included being held underwater, and having the hair of the animal that bit you rubbed on the wound.

Eventually, doctors figured out that they weren't curing anyone with these efforts and discovered vaccinations that were effective. But rabies bites were a big problem in those days. So were hairballs. Why they were combined is anyone's guess.

Giant Sea Monster Mystery

Have you ever heard of the Loch Ness monster? It is a large water creature that has long been a part of Scottish folklore, said to live in a lake in the Scottish Highlands. Nobody has real proof that this creature actually exists, and most of the scientific community regard the story as fake.

But there is another sea creature (100% real) that scientists are studying that is equally mysterious...the giant squid! This large ocean dweller can reach the size of 33-43 feet in length. But that's purely scientific estimations, because the giant squid is hardly ever seen alive. In fact, there are only three times when a live giant squid has ever been captured on video.

Scientists are learning about the squids through research of dead giant squid who wash up on the beach. But even then, there aren't that many that do that. There are only 12 giant squid remains that are on display at museums or aquariums in the entire world. They remain largely a mystery to marine biologists and researchers because they live AND die so far away from where we can see them. They live their lives in the deep, DEEP ocean.

Here's some things they do know about these big elusive swimmers. Giant squid seem to be all over the world, although they are not often found in tropical or polar climates. They likely feed on deep sea fish and smaller squid. Do they have many predators? The sperm whale and pilot whale seem to be the biggest threat to giant squid. Their remains have been found in whales' stomachs and some whales even show scar marks from battles with the squid.

If you ever ran across a giant squid in the ocean (don't worry, it's not very likely) here's what you might expect. Their body, or mantle, is around 7 feet long with tentacles, including two feeder tentacles which can reach 33 feet in length! These come in very handy for catching prey.

They have eyes the size of dinner plates. Yikes! That's creepy. And their brain is the shape of a donut.

Recently scientists have been able to sequence the genome of a giant squid. A genome is the complete set of genetic instructions for that particular squid. This means scientists are learning the genetic roadmap of the giant squid and will be able to compare it to other squid relatives. Then maybe they can figure out how the giant squid got so big, how they remain such a mystery, and other cool things about them. It seems that even the perplexing giant squid is no match for modern science and curious researchers.

The Forest Is on Fire? These Adorable Creatures Don't Care.

Pika are adorable little animals that live in rocky terrain, like mountainous areas. They are related to rabbits, but are a little smaller in appearance. Pikas live in small underground burrows.

Johanna Varner was researching pikas while working for her doctorate degree in college. She was studying several groups of pikas that live in different altitudes and how the altitudes and temperatures affect their living environment.

She was devastated to learn of the forest fires in 2011 that burned nearly 6000 acres of the Mount Hood National Forest, where she had been performing her research and carefully tracking the pika in the area. The area was 100 percent burned.

Not only did she feel like all of her hard work was gone, but she was also upset to think of what had happened to the pika. She had observed them gathering and storing food all summer long, and was devastated to think that they hadn't survived.

Johanna had put temperature sensors in many of the mountain crevices that the pika used as their underground burrows. She decided to collect her sensors before leaving the charred forest. She expected them to be mostly melted, but the sensors were still working normally. This gave her hope.

She took the sensors back home and was able to read the data they had been recording. It seemed that on the days that the forest fires swept through the area, the temperature in the underground burrows never got much hotter than a normal summer day. This gave her hope that the pika had been able to survive the forest fire.

This is the first research that suggests that sheltering in place during a forest fire is a viable option for some species. Johanna's research prior to the fire will also help scientists study how a species can come back and recover after a fire.

In the years following the fire, Johanna has followed along as the pika population in the area continues to grow back. Although it isn't as high as it was before the fire, we are just glad that some of those adorable little pika were able to survive!

How Good Manners Led to This Well Known Invention

Renè Laënnec was a French physician who understood lung troubles more than most doctors. He himself suffered from chronic lung disease and his mother died of tuberculosis when he was only 5 years old.

After the death of his mother, Renè was sent to live with his uncle who was the dean of a medical university. By the age of 14, he was assisting with sick and injured patients at the hospital and by the age of 18, he had been appointed a surgeon. Can you imagine? At only 18 years old!

With his uncle's encouragement, Renè got his degree in medicine and continued to be interested in medical research, as well as assisting patients.

Up until 1816, doctors diagnosed a patient's heart and lung trouble by listening directly to the heart and lungs. This meant that they would actually place their ear directly on the patient's chest and hope to be able to hear well. It wasn't a perfect practice. It didn't take much for the sounds to become muffled and difficult to interpret. If patients were larger, sometimes the sounds weren't apparent at all. And for women, it could be really awkward!

One day, Renè was having trouble hearing the heart of a young female patient displaying symptoms of heart disease. She was shy and he felt weird pressing his head against her chest. He remembered a simple lesson in acoustics. He rolled a piece of paper into a cylinder, placed one end to the woman's heart and lungs and the other end to his ear and was delighted to hear more clearly and distinctly than ever before.

The first instrument he built was a hollow wooden cylinder. He called it a stethoscope, taken from the Greek words stethos, which means chest, and skopein, which means to examine. He later tailored his design to include three detachable cylinders, meant for more practical use.

Renè became fascinated with studying lungs and put his stethoscope to good use. He is credited with determining normal and abnormal lung sounds and categorizing the different types of lung sounds of his sick patients.

Two hundred years later, we remember Renè Laënnec for thinking outside the box and not only being a pioneer in his field, but also a great doctor who just truly wanted to help his patients. The stethoscope is a worldwide tool used by practically every doctor in the world today. And to think it all started by rolling up a piece of paper to listen to a girl's heart.

Even the Most Brilliant Scientists Can Have the Most Bitter Enemies

There are lots of fantastic stories of fierce rivalries between scientists. Often, it either had to do with who was right and who was wrong or...who got the credit for some brilliant idea or invention.

For instance, Charles Darwin and Sir Francis Owen fought over who came up with the theory of evolution. This theory had to do with how animal species changed and adapted over time and it was groundbreaking stuff in the 1860s. Most historians believe that Sir Francis Owen was consumed with jealousy of Charles because Charles had gotten so famous because of his research.

Although their theories were very similar, Francis was always attacking Charles. He even helped edit an early copy of Charles' landmark book and told Charles to keep in phrases like "I think" or "I believe" instead of phrases where he stated something as absolute fact. Then after the book was published, Francis attacked him for using those words saying they were unscientific. That sure seems sneaky and rude!

Then there was the fight over who got credit for discovering calculus. Don't worry, I won't get too detailed here about calculus other than to say that it's an advanced form of mathematics. All the way back in the late 1600s, two guys actually came up with it at the same time! Sir Isaac Newton and the less well known Gottfried Wilhelm Leibniz. Isaac had written his papers on calculus in the 1660s but never published them. Gottfried published his work about calculus in 1684. I'm sure both papers were riveting!

But the two mathematicians each thought the other was a liar and a thief. Historians now believe that they both happened to discover the principles of calculus on their own. It's quite sad really that they had to go through so much stress over it now that we know that neither of the men were a liar or a thief.

Here's another from the late 1800s. It involves the brilliant mind of Louis Pasteur who was making all kinds of discoveries in microbiology and chemistry. Louis was French. An up and coming physician, Robert Koch of Germany, was also doing good work in these fields.

They were respectful of one another until...Louis disagreed with one of Robert's responses to his work. Louis, in defense of his work called Robert's contributions to science, "recueil allemand". Remember, these two guys didn't speak the same language. All "recueil allemand" means is "German works". That's not exactly insulting.

Unfortunately, Robert was told that these words meant "German arrogance".

Uh oh. That's not good. This simple misunderstanding turned into a scathing argument that led to the two brilliant men hating one another. They each went on to contribute greatly to the field of science and curing diseases but they never forgave each other. All scientists had to choose a side at the time which was really unfortunate for all of science.

There are so many more examples of these bitter scientific rivalries. I'll include only one more, but it's a doozy. This rivalry ended in a beheading!

One of the world's greatest minds in chemistry during the 1700s was Antoine Lavoisier. He is commonly called the father of chemistry, because so much of what he discovered has led to our understanding of chemistry today. He even discovered and came up with the names for both hydrogen and oxygen. He was a scientific powerhouse.

But there was another scientist in France at the time that Antoine did not agree with. His name was Jean-Paul Marat. Jean-Paul was a believer in the notion of "animal magnetism". Antoine made fun of him for believing in this and although it was mean, Antoine was right. There is no invisible force surrounding everything like there is in Star Wars. And that's basically what Jean-Paul so passionately believed in.

Jean-Paul did not handle Antoine's public shaming very well. He pestered Antoine for years afterward, constantly publishing insults of his work. Then the French Revolution happened and Jean-Paul became a key leader. There was an uprising and Jean-Paul was killed. His allies went after his longtime scientific rival even though he had nothing to do with his death.

More than 17,000 people had their heads chopped off during the French Revolution. That's both awful and gross, right? Jean-Paul's followers captured Antoine and put him to death by the guillotine as well. All because he let a scientific difference of opinion turn ugly.

Fortunately, even though scientists can still get pretty bent out of shape about disagreements with each other, it's been a long time since the guillotine was used to settle such a dispute.

Can Science Explain a Haunted Elsa Doll?

The purpose of this story is not to keep you awake at night or to scare you. It's actually the opposite. We're going to dive into the science behind ghost sightings so that the next time the hair on the back of your neck stands up and goosebumps appear on your arms, you can hopefully remind yourself of the science behind the spookiness.

But first...a ghostly doll?

I'm sure you've seen Elsa dolls that sing the song from Disney's ultra popular Frozen movie, maybe you've even had one. One family in Houston, Texas, had an interesting experience when they threw one away. Weeks later, they came home to find it sitting in their living room. Yikes! Everyone swore that they had nothing to do with its reappearance.

It was then that Elsa stopped singing her song when her button was pressed. Instead, she started speaking Spanish. The family didn't speak Spanish so nobody knew what she was saying. They double bagged her, and put creepy Elsa back in the trash.

The family then went on a trip and when they returned...you probably guessed it. Elsa was back. This time she was in the backyard. At this point, they mailed her to a friend in Minnesota who wasn't afraid of haunted dolls. He strapped Elsa to the front bumper of his truck and as far as anyone knows, that's where she still is today. I hope that guy is okay! I don't think I'd want her anywhere near me or my truck.

But what really causes these strange happenings and hauntings? Nearly everyone has a good ghost story. Can science help explain these strange encounters?

Fortunately, that answer is a definite yes. There are 3 main theories that explain ghostly experiences. They include electromagnetism, sounds waves, and our runaway imaginations. Let's dive in.

One theory that explains ghosts is electromagnetic fields, or EMFs. An EMF is when a field is created by an electromagnetically charged object. Small changes to these fields can give people the feeling of a paranormal experience by affecting the brain in a subtle way.

A neuroscientist in Canada, Michael Persinger, believes that EMFs account for a great deal of ghost sightings and experiences based on his studies. Not all scientists agree, but this is one explanation.

Another explanation was discovered by an electrical engineer who was working by himself late into the night. He was suddenly overcome with the idea that he was being watched and felt extremely uncomfortable. Then he thought he saw a ghost out of the corner of his eye. When he got up the nerve to look directly at it, the ghost disappeared. That's what ghosts do, right?

But then the engineer, Vic Tandy, discovered that something called infrasound was coming from an air conditioner in the room that he was in. Infrasound is a low frequency sound that we can't pick up with our ears. But our bodies detect it in other ways. Some infrasound can cause our eyes to vibrate. This could obviously make it seem as though we were seeing a ghost.

Vic went around to supposed haunted places all over England and discovered the presence of infrasound at nearly all of them. This makes his infrasound theory a plausible one for explaining supposed hauntings.

But what about the Elsa doll? I hear you. We haven't gotten to an explanation for that one yet, have we? Some scientists suggest that the principle of suggestibility is what causes ghostly experiences. This means that if we are expecting a certain place to be haunted, our awareness will pick up on things that wouldn't bother us ordinarily and we will blame things on ghosts.

I'm sure that suggestibility is the root cause for us thinking that spooky noises or sights are ghosts, but none of that explains the case of the haunted doll. Well darn. Sorry, I tried my best. If you come across a doll that you think might be haunted, mail it to Minnesota. They know how to handle these things up there.

The Scoop on the Super Soaker

Have you ever blasted somebody with a Super Soaker water gun? Or perhaps you were the one who got blasted? Either way, Super Soaker water guns have been a staple for summer fun for over 30 years. Did you know they were invented in a basement by a NASA engineer?

Lonnie Johnson was always super smart and curious as a kid. He loved experimenting and tinkering with building robots and engines. When he was a high school student, he entered a robot he had built named Linex, into a local science fair put on by the Junior Engineering Technical Society at the University of Alabama.

Linex was about three and half feet tall and could turn around and move all around on its wheels. He had shoulders, arms, and wrists that all swiveled. Lonnie won first place with his robot, which was a big deal at the time. Lonnie is Black and in 1968, the schools were still segregated. Black kids couldn't go to school with White kids. Lonnie's robot was the only entry from a Black student, but it was so good that no one could deny that Lonnie deserved first place.

Lonnie went on to college and was later recruited by the Jet Propulsion Laboratory at NASA. He was fiddling at home on one of his projects for them (a refrigeration nozzle) when he accidentally launched a watery stream across the room. He thought this would make a great toy. The idea of the water gun was born!

He decided to put the idea to the test. After his work day, Lonnie would come home and experiment on his idea in the basement. When his first prototype was complete, he handed it off to his 7 year old daughter to blast everyone in the neighborhood.

In order to get the water gun into large scale production and into stores, Lonnie needed to team up with a toy company. Larami was a toy company that showed some interest and they set up a meeting. "I took the gun out of my suitcase," Lonnie remembered. "They asked if it worked and I shot water across the conference room."

Larami helped launch the water gun, then called the Power Drencher, in 1990. Everyone loved it. They did some tinkering on it and changed the name to the Super Soaker and sold more than 2 million of them in the summer of 1991.

The toy water gun has earned well over one billion dollars in the last 30 years and in 2015 it was added to the National Toy Hall of Fame. Thanks to this fun invention, Lonnie is now worth over $300 million dollars. Lonnie Johnson's story is a great reminder to follow your dreams and stay curious. And never pass up a chance to squirt someone with water!

Take This Snow and Shovel It

What is called white gold and has the ability to make or break somebody's holiday weekend? Snow! The cold, fluffy powder is an absolute staple for the downhill snow skiing tourism industry. The lack of snow can shut down an entire operation and makes lots of skiers pretty sad. But what can you do when you are dealing with Mother Nature?

In 1934, the Toronto Ski Club had a big ski jumping competition planned. But the weather didn't cooperate. Instead of canceling due to lack of snow, the ski club asked for help from the University of Toronto. They used the ice from the university's ice skating rink, and made it into shaved ice to take the place of snow. Not only did it sufficiently coat the ski hill and jumping area, the jumps the skiers made that day were even longer than usual. It was an impressive substitute and got people thinking.

During the early to mid 1900s downhill skiing was a hugely popular activity that created a large tourism industry. But the downside was that it was so dependent on relying on nature. Not only could there be a lack of snow, but the air temperature could affect the quality of the snow and a quick rain could make the ski slopes turn icy.

In 1947, one of the first large ski resorts was facing a major problem. It was Christmas and there was no snow on their ski runs. Over New Year's weekend, the owner rented a wood chipper, bought 450 tons of ice and got to work. He covered his most popular ski hills in shaved ice and by the end of the weekend, he had financially broken even. But the task was labor intensive and depended on finding an incredible amount of ice. Luckily, three engineers who were avid ski fans got to work on a plan.

The first two attempts at creating snow had used ice as a substitute, but these three engineers thought about using two key components to natural snow. These were water and cold air. Cold air was readily available in the winter. So they invented a machine using pressurized air and pressurized water. When the cold air hit the water, it broke the water droplets into even smaller particles, allowing them to freeze quickly. Thus creating "snow."

These first ingenious attempts at creating snow gave many skiers a chance to enjoy their favorite activity and allowed the workers in the tourism industry economic success. Nowadays, all ski runs from vast ski resorts to small ski hills, use this technology to extend their season and make a snowy run whenever Mother Nature isn't cooperating.

Cruising In a Taxi with NO Driver?!

In the not-so-distant future, you could hail a taxi or catch a ride and the car that picks you up could be completely autonomous...meaning NO driver!

While it seems like this is something futuristic that you might see in a movie, the idea is not that far off. Driving companies have teamed up with automotive manufacturers and have created several different fleets of cars that operate completely off of computer technology. They've been working on the technology for several years, and it looks like launching to the general public may not be too far away.

The systems that these cars use are full of multi layered sensors. The job of the sensors is to keep track of all people and objects, no matter how hidden, dark, rainy, or foggy it might be. This means that unlike humans, who can only see what is directly in front of them, these robocars should be able to sense what is going on ALL the way around them.

They also run off of electricity, which means that they don't use gas. This makes them better for the environment. These cars may look a little different, too. Okay, some may look a LOT different! If there are no drivers, there is no need for a front seat. There also doesn't need to be a steering wheel or brake or gas pedals. So many of the things that take up space in a normal car aren't needed when the car is electric and runs 100% by a computer. So that means more space for seats and legroom.

The companies working on these vehicles hope that the cars will be an effective ride-share or car pooling option for big cities. It will be a way for people to save money, time, and the environment. In fact, a company in San Francisco has been using their robocars for several years as an option for their employees to get to work. This allows them to have their cars practice navigating around a big city, while they work out any kinks in their system. And it gives their employees a stress-free way to get to work. That's as long as they don't get stressed about giving a robot car complete control of their safety. This will probably take some time for some of us to get used to.

So why aren't these robocars out in full force right now? Well, the companies and manufacturers are still going over every detail to be sure they are safe. In theory, the computer system would have less error than human drivers. Human drivers certainly make LOTS of mistakes when driving. This includes driving while sleepy or distracted, not looking before a turn, or just having delayed reactions. In some big cities, these vehicles are out being tested. Perhaps they are driving employees to work, or they are out making grocery deliveries, or driving robocar designers around looking for ways to make the technology better.

All the miles these first robo cars are logging are helping the design teams make the cars better at their jobs and able to handle difficult situations. That's because when it's you in one of these cars, you want to make sure it's super safe!

Science Fair Project Underwhelms Judges, but Impresses Scientists!

Simon Kaschock-Marenda was a 6th grader in need of a science fair idea. His parents had recently tried to cut sugar out of their diet. Finding this interesting, Simon decided to test feeding different types of sugars and sweeteners to fruit flies to see the results.

He kept the vials of fruit flies on a shelf in his closet and put different sweeteners in each vial. The results were eye opening.

The fruit flies that ate the sweetener Truvia, all died within 6 days. Simon's dad is a biologist and he thought maybe the project had gotten contaminated. So Simon and his dad repeated the project in his dad's laboratory, where all the variables could be controlled. The results were the same!

Simon turned in all of his research for his science fair project, but it turns out that the judges weren't too impressed. He didn't even win a ribbon for his work!

But, while he may have not impressed the judges, researchers found the results fascinating. A full fledged study was done. The results were that Truvia, which is made from the Stevia plant and contains erythritol, seems to be problematic for the fruit flies. Erythritol has been studied and is safe for humans to eat and is even found naturally in some melons and grapes.

The fact that it seems to affect fruit flies has scientists excited that perhaps the erythritol can be used as a natural pesticide. Further studies will be done to see if this will be useful and also if the erythritol will have similar results with other flies and bugs. Having a natural pesticide that will deter flies and other insects from eating and disturbing crops will be a very important discovery for agriculture!

Results of the first study were published and listed Simon as an author of the study. He thinks that is pretty cool and so do the teachers at his school. Even if he wasn't rewarded for it at the science fair!

The Dancing Lights

The sun and the earth are 93 million miles apart. That's a long way! But each year teeny, tiny particles of the sun blow toward earth in a solar wind and create a brilliant light show that has been seen for thousands of years.

The aurora borealis, also known as the northern lights, are a dazzling, natural light show that can be seen in many northern countries around the globe. Beautiful bright lights in all shades of green, pink, purple, yellow, red, and blue shimmer and dance against the night sky in different patterns each night.

So how does this happen and what does the sun have to do with it? The sun produces electrons and protons which are tiny little energy building blocks that make up pretty much everything. Some of the electrons and protons escape from the sun and are blown out into the atmosphere by solar winds (winds around the sun). Some travel all the way to earth, but then they are stopped. Did you know the earth has a magnetic field all the way around it, acting like a shield? The magnetic field forces these electrons to travel around earth. So they essentially hit earth in the middle and then are forced either north or south.

As they reach the north or south poles, some of the solar winds and thus the electrons and protons come into contact with the earth's atmosphere. The atmosphere is like a blanket of gas that protects the earth and allows us to breathe. When these protons and electrons mix with the gas in the atmosphere something beautiful happens. The energy from these particles is released in the form of light. All of these particles releasing their energy together creates an incredible light show with a rainbow of colors.

So where can you see these lights? The northern lights or aurora borealis can be seen best in countries in the northern part of the northern hemisphere. Sweden, Finland, northern Canada, and Iceland are all well known for their displays of northern lights. In the southern hemisphere you can see the southern lights or aurora australis. These lights are not quite as bright and easy to see, but your best chance of viewing them would be Antarctica or southern Argentina.

The best time to catch the lights would be between December and April, but you'll have to stay up late. The best time to see the most dazzling lights is on a very clear night between 10pm and 2am. And you'll want to make sure you are in a very dark place, away from the lights of the city, which would make the show less intense.

People have been studying this natural phenomenon for a very long time. There are cave paintings showing this display of light that are over 30,000 years old. Many civilizations had their own folklore for the light's meaning. Vikings believed that the lights were reflections from the shields of the Valkyrie, their immortal female warriors. Ancient Finnish legend tells of a small arctic fox that ran so fast across the snow that sparks flew into the night sky. The Minominee Indians believed the lights were from the torches of giants that lived far away.

If you ever get the chance to see the lights you should definitely do it. And now you will be equipped with the knowledge of how tiny energy particles from the sun are responsible for creating this radiant display here on earth.

Happy Cows

It makes sense that happy cows would produce more milk, right? Almost everyone is more productive when they are happy. So what makes cows happy? Green grass, mild breezes, warm sunshine...

So you can imagine that those three things are probably in short supply in a country like Turkey in the middle of winter. So what's a farmer to do?

Izzet Kocak is a Turkish dairy farmer who loves his cows. But he has seen a decrease in their milk production during the winter. Unfortunately, feed prices are on the rise and Izzet needs to produce milk to feed his family and his cows. He knows that happy cows are good producers. In past years he has played classical music to his cows, who have to spend much time in the barn due to cold weather conditions. But he was looking for other alternatives.

Izzet had seen a Russian study where cows were fitted with virtual reality headsets to simulate being in a nice green pasture. He decided to perform an experiment on his own cows. Two of his cows were fitted with virtual reality headsets. These cows got to experience that they were in a sunny pasture on a spring day.

Izzet observed the results for ten days. The outcome? The cows outfitted with the headsets had been producing 5.8 gallons of milk per day. With the virtual reality goggles, the cows started to produce 7 gallons per day! Izzet said the quality of the milk was also better. He plans to order 10 more headsets.

This is an interesting study where technology can be used to help farmers with their production. Of course, it raises questions about how ethical it is for the treatment of cows. Some people fear that it could be taken too far. Cows may live their whole lives in some sort of pretend universe where they think they are outside, when they are actually trapped in a small barn. I hope Izzet's cows continue to enjoy simulated sunny days until spring when they can return to their actual pastures.

Don't Be Blue

Imagine doing research for magnetic materials to be used on computer hard drives and stumbling across a new pigment for...the color blue?

That's exactly what happened to Mas Subramanian when he was working in his lab one day. Mas was looking for materials that had magnetic properties that could be used in computers and other electronics. He would mix up different chemicals and bake them in a furnace at over 2400 degrees. When he pulled out one particular mix he noticed a change. Adding some manganese oxide to the batch had produced something extraordinary.

"I was shocked because the samples came out so blue," he said. "At the first moment I (knew) it was going to be an amazing discovery."

Blue is one of the trickiest colors to create on earth. The pigments used to create the color blue are very rare. In fact for a long time, the minerals used to create the color blue were more valuable than gold! About three hundred years ago, humans found a few basic recipes to create the color and most color development for blue stopped. Until Mas had his happy lab accident.

This is the first new blue color pigment to be discovered in over 100 years and interestingly, one of the most dynamic and brilliant forms of blue ever created.

Any color, whether it is blue, green, or red, is produced using pigments or dyes. Dyes come from organic material, and are typically used in foods and clothing.

Pigments are usually made from inorganic material like minerals and rocks. Sometimes they are harder to come by, but they hold their color a lot better and last much longer. Mas's new color was one of these pigments. He named it YInMn.

It seemed that the secret to YInMn being such a brilliant blue was the shape of its molecules. They are arranged in a triangular pyramid shaped structure which is an uncommon shape in minerals.

As well as being a vivid and brilliant color of blue, YInMn is also very durable, non toxic, and highly reflective. It reflects infrared radiation so it will likely help keep vehicles and buildings cooler. Mas has begun working with paint companies and energy conservation groups to use his new finding to produce helpful materials for the future.

The Shark Tooth That Spent 25 Years Inside a Surfer's Foot

Twenty five years ago, Jeff Weakley was just another college kid hanging out at the beach, surfing. That is, until something grabbed his foot in the water and bit him! Luckily, it was a pretty minor bite but he did have to wrap it and use crutches for a while. Although he didn't see who tried to take a bite out of him, Jeff and everybody else assumed that a shark was responsible.

Fast forward 25 years. Jeff gets a blister on his foot that stays irritated. He decides to open the blister to see if it will relieve the pressure and he is shocked to find a tooth fragment in his foot! It only makes sense that this tooth belonged to the predator that got him so many years ago.

Jeff thought he might make a necklace with the tooth as a fun story to tell, but then he read about scientists analyzing DNA from shark teeth. He decided to donate the tooth to science and see what they could learn. He admits that he was a little hesitant...what if it just turned out that he was bitten by a mackerel or some other less threatening fish?! Then his story would be a lot less fun.

The researchers at the Florida Museum of Natural History were excited to perform the study, but they were skeptical that they would be able to extract enough

DNA to be able to determine the type of fish. I mean, the tooth had been inside Jeff's foot for 25 years!

Surprisingly, the scientists were able to clean the tooth, remove the enamel, and extract DNA from the tissue inside the tooth cavity. They were able to piece together enough of the genome sequence needed to identify the type of shark that had munched on Jeff's foot. The result? 25 years earlier, Jeff had been bitten by...a blacktip shark!

Blacktip sharks are responsible for quite a few of the shark bites in Florida. While these sharks usually don't attack people, sometimes hunting mistakes cause them to bite a human by mistake when they are pursuing schools of fish.

Science is amazing, right? After 25 years, scientific advancements were able to help Jeff identify the exact breed of shark that bit his foot. Jeff is grateful that the scientists at the Florida Museum of Natural History were able to provide him with the missing piece of his story. It's cool that he knows a little more about what got him all those years ago. And he doesn't hold any hard feelings toward blacktip sharks. In fact, he said he has eaten blacktip shark for dinner a few times and it was delicious.

The Power of Hope

Hope is a feeling of expectation for a certain thing to happen. Hope is a great motivator for people. It pushes you through the hard stuff and helps you persevere when things get tough.

In the 1950s, biologist Curt Richer performed an experiment that proved just how valuable hope is. Although parts of the experiment were gruesome and cruel, the results really do prove that hope is a powerful motivator.

Curt performed his experiment on rats. To begin, he took 12 rats that were raised domestically, meaning they were used to being handled by humans. He placed these rats in individual buckets of water and observed how they handled it. Three of the rats drowned within 2 minutes of hitting the water, but in a surprising twist, the other nine swam for days before they drowned.

Next, Curt took 34 wild rats, known for their swimming capabilities, and did the same experiment. Since these rats were tenacious, wild rats, he expected them to swim for a long time. The results surprised him. Every single wild rat drowned within a few minutes of being put in the water.

So what was the difference that kept the tame rats swimming while the wild rats gave up so quickly? Curt predicted that it was hope. The tame rats had been fed and cared for by humans. The wild rats perceived the situation as hopeless.

Curt tweaked his experiment one more time. He took a group of similar, wild rats and placed them in the water. Then just before they were about to succumb to drowning, he pulled them out of the water . He held them and helped them recover. Later when they were placed back in the water, Curt found that these rats kept swimming. They didn't give up and go under. They had a reason to keep swimming since they now had hope that they would eventually be rescued.

Although it's very sad that so many rats had to suffer and die in his experiments, Curt really showed the power that hope can have for both humans and animals. So the next time you are in a situation that has you struggling, find something to be hopeful about. And in the words of Dory, the little fish from Finding Nemo, "Just keep swimming!"

The 17 Year Old Who Worked on the World's First Calculator

I'm betting you've used a computer and probably a calculator. These machines make learning incredibly easy and fun. Calculators help us solve difficult math problems with ease and computers help us find out anything we want, right away.

What if I told you that one of the first computer programmers was a young girl who was fascinated with math? Ada Lovelace was only 17 years old when she met her mentor and began to help with the development of tools and ideas that were revolutionary for their time.

Ada was born to a famous poet father who left her and her mother when she was just a baby. Her mother believed that Ada should have a great education and so she had excellent tutors. When she was 17, her tutor introduced her to Charles Babbage.

Charles was a mathematician and inventor. He created the Difference Engine, which is sometimes thought of as the first calculator. Ada was fascinated with it and Charles thought Ada was very clever. He became her mentor and together they worked on many projects.

One of the projects that Charles was working on was the Analytical Engine, which would be even more complex that his Difference Engine. While working with Ada, he asked her to translate some information from a French engineer. Ada did the translating, but also wrote in a bunch of her own notes and ideas on how the machine would work.

She likened it to a weaving machine that could form its own complex designs. She thought the machine could use patterns or codes to work with both numbers and letters. Her notes and theories have earned her the title as one of the first computer programmers. She was coming up with these ideas before computers ever existed.

Sadly, Ada died of cancer at the age of 36, way before her notes and brilliance would actually be discovered. In 1979, the United States Department of Defense honored her memory by naming a new computer language they had developed, Ada.

All Roads Lead to Rome

The Roman Empire ruled for nearly 1000 years and had a vast expanse of land covering much of Europe and parts of northern Africa. The empire's capital and early beginnings centered around Rome (current day Italy). It was known for its military prowess, and social and political structures that continued to shape society long after its fall. There are many reasons that the Roman Empire was so powerful. What would you think if I told you that roads were one of the major reasons?

The Roman Empire had over 250,000 miles of roads! That's a lot of walking since they didn't have cars back then. They started building these roads over 2300 years ago. This was before bulldozers and dump trucks, yet these roads were incredibly well engineered. Over 50,000 miles of these roads were even paved with stone. Some of these roads are still used today!

So, how were these roads even built if they didn't have heavy machinery? Road engineers decided where the road should be placed. Then, surveyors used a rod and groma to make perfect right angles. Rods were placed in a grid pattern and road crews got to work digging. So much digging! Many of these roads were dug down three feet deep to establish the firmest ground. Then sand and small rocks were packed in the area to allow proper drainage when it rained. Finally, if the road was to be a paved road, it would be fitted with specially shaped paving stones and filled with a limestone concrete that the Romans had invented. All roads were built with a crown. No, not a shiny gold headpiece. A crown was a specific technique where the road was angled a very slight amount to allow for water to run off of it.

These roads went everywhere. And they were built to be the most efficient for the traveler. That means they didn't go around rivers or mountains. The roads were built straight over and through them. So it was really an impressive feat that these engineers had to pull off. They had to build bridges and tunnels to accommodate certain roadways.

Most of the roads were outfitted with mile markers, so travelers could measure their distance. Every 16-19 miles, inns, taverns, wagon and wheel specialists popped up with little towns growing all around them. A mail service was started because road travel made it easy to deliver messages.

As the roads and the towns grew, so did the Roman military. With the roads it was easy for the military to travel, meaning they were able to conquer even more lands. The Roman Empire continued to grow. Some of the roads built in the United Kingdom are still in use today.

You can see how these extraordinary roads were really the backbone of the success of an entire civilization that ruled for over 1000 years. We may take roads for granted today, but until modern times, no other civilization in history had as impressive or extensive of a road network as the Roman Empire.

The Girl Who Discovered Underwater Mountains

Marie Tharp created the first map of the ocean floor. But get this. She wasn't allowed on the research boats because she was a woman.

Marie was born in 1920. Her dad was a surveyor, collecting soil and rock samples, and making charts of specific locations. Her family moved around a lot. By the time she graduated high school, Marie had attended 17 different schools!

She was smart and fascinated with rocks and map making. But back then it was not common for girls to enter scientific careers. Most women became teachers, nurses, or secretaries. A stroke of luck allowed her to pursue her passion. It was during World War II, and with so many men being involved in the war, Marie got the opportunity to get her degree in geology.

Marie also loved studying the oceans, marine life, and ecosystems. She and her research partner, Bruce Heezen, decided to make a map of the ocean floor. Back then, everyone thought the ocean floor was likely flat and made of mud. They couldn't have been more wrong!

In order to make the map, they needed to collect data. And because she was a woman, Marie wasn't allowed on the research boats. There was a widespread superstition that having women on the research boats would bring bad luck. Bruce gathered the information using SONAR measurements of the ocean's depths. The material was sent back home to Marie who did all of the charting. Using just a pen and a ruler, Marie drew out all the data.

Marie and Bruce's work showed that the oceans weren't flat. They were actually filled with canyons and mountain ranges, just like the land. Marie charted all of the oceans on earth and in 1977, published a map called, The World Ocean Floor.

Marie's work helped to prove a theory that had originally been dismissed as crazy. In 1912, Alfred Wegener, a German meteorologist had proposed the theory of continental drift, the possibility that continents and land move over time. This idea was considered preposterous! Everyone thought Wegener was crazy and forgot about it. But the more Marie worked on charting the ocean floor, the more she thought it was possible. She charted 40,000 miles of ridges on the ocean floor. She was able to see the impact that small earthquakes had in the areas she thought were shifting. All over the globe she plotted these ridges where earthquakes seemed to occur.

It took many years of work for her theories to be accepted as fact. Most people in the scientific community didn't believe it. Jacque Cousteau, the famous oceanographer, wanted to disapprove Marie's theory. He even went out in his boat and used underwater video cameras to record the ocean floor.

But his recordings just proved that Marie's theories were correct!

Eventually, the geologic community began to accept that Marie's findings were accurate. As a woman, she often didn't get the credit she deserved for her work. Today, we see her as one of the world's most significant mapmakers who stood tall during a time when women were expected to stay in their proper place.

Nice Shades, Bro!

Take one guess at one of the most ingenious inventions for those living in the Arctic thousands of years ago. If you said fire, you're not wrong. That one is pretty important to everyone. But think again. What about sunglasses?

Sunglasses? Aren't those kind of a fashion accessory? Yes, and while today they look fashionable, back then, the first sunglasses were created out of absolute necessity.

When the sun hits the snow it is highly reflective, and the Arctic has a LOT of snow. When all of this ultraviolet light hits the eye, it can cause a condition called photokeratitis. This basically means that your eye gets a sunburn. Ouch! The damage to the cornea can cause temporary blindness.

This condition became known as snow blindness to the indigenous people living in the Arctic. It was painful and obviously dangerous, to be temporarily without the use of your eyes. So people invented the very first sunglasses. They were carved out of wood or ivory, with laces to tie behind the head. Small, long slits form the eye holes to see out.

This prevented the wearer of these glasses from having to squint their eyes. And it shielded them from the majority of the sun's rays.

Hunters would wear them to protect their eyes while hunting or whenever the sun was particularly bright. I'll bet the original inventors of the sunglasses would never guess what a fun and stylish accessory they would be so many years later!

Can You Really Just Grow a New Ear...on Your Arm?

Hopefully this book has opened your eyes to all of the crazy possibilities that science and technology are capable of creating. Some of these stories have shown you just how far people have come from seemingly simple devices like the stethoscope to all of the technology we use on a daily basis. Humans are constantly thinking outside the box.

So here is a story that shows just a glimpse of what people can expect in the future. New scientific developments will be able to help all kinds of patients, sometimes in unexpected ways.

Army Private Shamika Burrage was involved in a terrible car accident when she was just 19 years old. She suffered head and spinal injuries, but perhaps the hardest to cope with was the loss of her left ear. It was so badly damaged that doctor's had to remove it.

Shamika struggled with the loss. She could still hear out of her right ear, but she felt apprehensive about the way she looked. She went to counseling to try to cope with her feelings. Her counselor encouraged her to talk to a plastic surgeon. This led to a conversation that would change her life and the lives of others struggling with similar situations.

The surgical team at William Beaumont Army Medical Center had BIG ideas. They wanted to get Shamika a new ear. Actually they wanted to GROW Shamika a new ear. She would actually grow the ear herself...on her arm!

The doctors collected cartilage (that's the flexible connective tissue in the body) from Shamika's ribs. The cartilage was shaped into an ear. The doctors then opened a flap of skin in Shamika's forearm and inserted the shaped ear. Because the ear was made out of tissue from her own body, the body accepted this new feature and the skin grew and closed around it. Blood vessels, arteries and nerves all linked into the new ear and it came to life. Within a few months the new ear was ready to be transferred to Shamika's head.

The surgery was a success. The doctors were even able to open her ear canal so that Shamika regained full hearing in both ears. She has sensation and feeling in the new ear, too. The goal is that people won't even be able to recognize that she had the procedure done because it looks so natural.

This was the first time that the Army doctors have performed this type of procedure but it has been done in other parts of the world. A doctor in China performed a similar surgery for the first time in 2017 and since then has used the procedure to help over 500 kids struggling from accidents or birth defects. Using our own bodies to create the things we need will open a lot of possibilities to help patients in the future!

Is There Not a Nobel Prize for Math Because Alfred Nobel Hated Mathematicians?

Have you heard of the Nobel Prize? It's a very prestigious award given to some of the biggest intellectual achievements every year. Alfred Nobel established these awards in his will in 1895, leaving a great sum of money behind to fund the awards. The first awards were given in 1901, five years after Nobel died.

In his instructions to set up the awards, Alfred stated that the prizes should be awarded "to those who, during the preceding year, shall have conferred the greatest benefit on mankind." He chose to award the areas of physics, chemistry, literature, medicine, and peace.

Hmmmm.... Notice anything important that Alfred didn't include in his prize categories? Math! Mathematics has played a pretty crucial role in the development of the world and humanity. Why do you think it wasn't included?

Of course, rumors started. Some people claimed that Alfred didn't award a Nobel Prize in mathematics because he hated math. Another rumor claimed that he had a rivalry with Swedish mathematician Gosta Mittag-Leffler. One rumor was that Gosta already had such an enormous fortune that Alfred didn't want him to be awarded any more money. Another rumor spread that Alfred didn't want Gosta to ever win one of his prestigious awards because they had a rivalry involving a girl. This particular rumor even claimed that Gosta had stolen Alfred's wife.

But digging into these claims, we discover that Alfred wasn't even married! Most historical sources say that while it is likely that Alfred and Gosta knew each other, there weren't any well documented rivalries between the two.

The most likely reason that there is no Nobel Prize in mathematics? It's a lot more boring than an intense rivalry. Most historians feel that Alfred just didn't think of offering a Nobel Prize for mathematics because he just wasn't that interested in math. Not that he hated it, but he wasn't as excited about mathematical developments as he was about science and literature.

So, what's a mathematician to do? They deserve some credit, too. Mathematicians, fear not! In 1936, John Charles Fields established the Fields Medal. It has been described as the Nobel Prize of mathematics and is awarded annually to 2-4 mathematicians under 40 years old who have made outstanding contributions to the field of math. The Fields Medal is known to be the top math award worldwide.

An Invention Born From Kitchen Clumsiness

Josephine Knight Dickson sometimes struggled in the kitchen. No matter how hard she tried, it seemed she was always getting small cuts or minor burns. If she didn't wrap it, the wound would be susceptible to infection and be a little painful. But her only options for wrapping were large and cumbersome, not to mention nearly impossible to apply on her own.

What's a girl to do? Her husband, Earle Dickson, worked for the company Johnson & Johnson, which made both surgical tape and gauze. Earle figured he could use these tools and devise some way to help his wife. He unrolled a length of the surgical tape and laid a small strip of the gauze over the sticky side of the tape. He then layered the tape with crinolin, which would allow the tape to be rerolled without sticking to itself. Josephine could then unroll a small bit of the tape, cut it off and apply it to her wounds when she had a clumsy moment.

Johnson and Johnson saw the potential when Earle showed his creation to the company leaders. A bandage that could be applied by a single person clearly had its benefits. However, it took a while before this new invention took off. It could be because the original Band-Aids were sold in 2 ½ inch by 18 inch strips, an awkward size for sure! But in the 1920s, Band-Aids were put into Boy Scout first aid kits and the rest is history. Families recognized the benefit of this clever invention and the sales took off.

Earle (with his wife Josephine credited for helping him) was inducted into the National Inventors Hall of Fame in 2017. This just goes to show that even solutions to a small inconvenience can help out in a big way and be hugely successful.

Earthquakes & Moonquakes

The ground below you quivers slightly and the pictures hanging on the wall rattle in their frames. What is happening?! It could be an earthquake!

Earthquakes are happening all the time all over the earth. The National Earthquake Information Center estimates that there are about 20,000 earthquakes globally each year, so about 55 a day. Of course, not all earthquakes are big earthquakes that rattle the walls or knock down buildings. Most earthquakes are very small and many can't even be felt at all.

People that live in areas affected by hurricanes and tornadoes know that these natural disasters usually have seasons, or times of year when they are most likely to occur. What about earthquakes? Earthquakes can occur at any time, day or night, sun or rain, hot or cold.

While earthquakes may not have a season, there are areas that are more likely for them to occur. This has less to do with climate and more to do with what is going on underneath the earth. The earth is made of a bunch of large pieces of rock, called tectonic plates. The plates are put together like a puzzle to form the earth. The place where two tectonic plates meet is called a fault. When two plates rub together at the fault lines this creates energy called seismic waves. This energy comes through the surface of the earth and creates an earthquake. Areas of the earth's surface that are directly over top of fault lines are more likely to have earthquakes.

Earthquakes are measured by seismographs that record the amount of shifting or shaking that occurs during an earthquake. The findings of the seismographs are then evaluated to determine the size of the earthquake. This is measured using the Richter Scale. The smaller the number on the Richter Scale, the smaller the earthquake. An earthquake that measures 1 or 2 on the Richter scale may not even be felt, while a 6 would cause substantial damage.

The largest earthquake ever recorded was in Chile on May 22, 1960. It measured 9.5 out of 10 on the Richter Scale. This earthquake was so massive that it destroyed the homes of 2 million people and caused shaking along 621 miles of coast.

The country of Chile actually expanded and got bigger thanks to the shifting of plates deep below the earth's surface. Whole towns shifted violently and moved 30 feet to the west in a matter of seconds. The next day, Chile was bigger by over 1,500 football fields thanks to the shifting of the tectonic plates.

Chile is a coastal country running along the Pacific Ocean on the western side of South America. When earthquakes happen near the ocean, they can cause tsunamis or big waves that can cause lots of destruction. These waves not only hit Chile with much devastation but even 12 hours later were hitting Japan after striking Hawaii and California. People had to run for the hills, climb trees, or hide out on rooftops while the flood waters caused by the tsunami swirled below them.

So what do you do if you are in an earthquake? Get somewhere safe, fast! The safest place would be in a field with nothing around to fall on you. If you are inside, get under something sturdy, like a desk or a table and protect your head. Stay put until the shaking stops. And if you're near the ocean, get to higher ground quickly if the earthquake is a powerful one.

There are even earthquakes on the moon and other planets. Moonquakes are usually less severe than earthquakes, but can last for up to an hour! Scientists think that many moonquakes have to do with the gravitational pull of the moon from the earth.

You Scratch My Back, I'll Scratch Yours

Do you know what a symbiotic relationship is in the animal world? It's a close relationship between two different species. These relationships help the species survive in the wilderness, making life easier for one or both creatures. In a mutual symbiotic relationship, both species benefit.

Just like we rely on our family and friends to help us and make life better for us, other living organisms do the same. There are some fascinating examples out there, so let's take a look!

The Nile crocodile is one of the largest reptiles on the planet and can reach 16 feet long and weigh 500 pounds. They live in the freshwater bodies of water in Africa. Although the crocodile mainly eats fish, it will attack anything that crosses its path, even zebras or small hippos. It doesn't seem like it would have any friends, and yet there is one creature that it will welcome into its domain.

The Egyptian Plover is a small shorebird that lives mainly on the sandbars of rivers in Africa. The birds are around 7-8 inches tall. A Nile crocodile will lounge on the sandbar with its mouth open. The Egyptian Plover will fly right into the croc's mouth and begin picking and feasting on the leftovers in the crocodile's teeth. The bird gets a free meal and the crocodile gets a quick dental cleaning. How's that for a happy story?!

Another tale of unlikely African animal friends is the warthog and the mongoose. Warthogs are fierce fighters and they can wreak havoc with their razor-sharp tusks. And yet, when it sees a pack of mongoose, it will lay down and allow the small fuzzy creatures to come and feast on the ticks and bugs in the warthog's coat. The mongoose gets a free, easy dinner and the warthog gets a thorough cleaning and massage.

Another fun example comes from the rainforest in South America. Most frogs in the rainforest are preyed upon by the snakes and spiders that dwell all around. But one tiny frog, measuring less than 3 centimeters, often cohabitates with the mighty tarantula. The microhylid frog has been known to live in the same burrow as the tarantula spider. The frog gets the protection of his roommate and in return he eats the small bugs that often try to eat the tarantula's eggs. What an unlikely pair!

There are tons of interesting relationships between different species. If this is something that interests you, it will definitely be worth your time to do some further research.

My Snack Melted!

Many important inventions and discoveries were found by accident which means that it's very important for scientists and researchers to always be paying attention and thinking outside the box.

Percy Spencer was a man who didn't see challenges, he just saw problems waiting to be solved. Born in rural Maine, Percy didn't have a lot of formal schooling and was mostly self taught. His curiosity and problem solving skills led him to become an engineer and land a job at Raytheon Manufacturing Company. This company is still around today. They make missiles, electronic warfare products, and military training devices.

In the early 1940s, Raytheon was working to improve radar technology to help with World War II efforts. And Percy was one of the best guys on the project. He had a knack for finding simple solutions to problems that seemed complex.

Percy was working to improve radar magnetrons. What are radar magnetrons? They are like an electronic alert that produces vibrating electromagnetic waves, or microwaves. He was trying to improve the power level that was to be used in radar sets.

Here is where fate intervened. After several hours of working on his radar magnetrons, Spencer reached into his pocket for his snack. But he found that his peanut cluster was all melted! Percy said, "It was a sticky, gooey mess!"

When this tale is told, many times a chocolate bar is in the story instead of a peanut cluster, but Percy Spencer's grandson, Rod, insists that the story has just been changed over the years. Rod says that his grandfather loved nature and would always carry a peanut cluster in his pocket to snack on and feed any squirrels or chipmunks that happened to be around. This is important to the story because chocolate would have melted at a much lower temperature than a peanut cluster, meaning the "microwaves" Percy produced that day were much stronger.

Percy was curious, naturally, so he put an egg under the tube he was working on. It exploded! He got some popcorn kernels and popped them, sharing popcorn with all of the staff. The idea of the microwave had been born.

It took nearly 20 years for the modern microwave to be developed and mass produced. The first microwaves were nearly 5 feet tall and cost thousands of dollars. But in 1967, the mass produced microwaves became an instant seller and now are in 90 percent of homes in America. But whether you are heating up popcorn, soup, or peanut clusters, we sure love how fast and easy microwaves make our lives.

A Superhero Robot to the Rescue

Around 350 natural disasters, such as hurricanes, tornadoes, and earthquakes happen in the world every year. And an estimated 90,000 people die every year because of these disasters. One of the reasons people die is because they get trapped in the debris of a fallen down house or building and can't get out. Search and rescue teams consisting of humans and dogs are great at doing what they can, but what if there was a superhero robot who could join the search party as well?

That may soon be an option. A robotics team in Italy has been working for 15 years on a robot that can aid in search and rescue missions. The robot, called iCub, stands just under 3 1/2 feet tall and even has a jetpack that it can use to propel itself out of difficult terrain. The robot has been designed with armor similar to that of comic book character, Iron Man.

iCub can walk, crawl on all fours, sit and use its "hands" to manipulate pieces of rubble. Humans can navigate the robot from a safe remote area, allowing the robot to explore the disaster area. This allows humans to investigate the area, note if anyone needs help and if the robot can't do it, the search and rescue teams can be better equipped to safely perform a rescue. The robotics team thinks that with its small size and propulsion capabilities, it might be able to go into areas that humans can't go.

The iCub even has a face, arms, and legs. The face is a little on the creepy side and looks like a doll. But if my house has fallen down on top of me, I'd sure be glad to see the iCub coming to the rescue.

If you are ever in a natural disaster situation, I'm sure you would be happy to be rescued by anyone.... but it's super cool to think you might be rescued by a superhero robot!

A Life Saving Heart Device Inspired by a Sailboat

It was the late 1960s in California. Hippies were running all over the place and everyone was fascinated with the Beatles and the Summer of Love. But in Santa Monica, California, on one particularly pretty day, an idea was born that would save thousands of lives.

Jeremy Swan was at the beach with his kids. It was pretty, but it was not a windy day. The air was still. Despite this, out in the bay, the sailboats were moving around just fine. Jeremy was impressed that they were able to make such good progress even without much wind.

That's when the idea hit him like a ton of bricks. Could this be the answer that medical science had been looking for?

Jeremy was born in a little town by the coast of Ireland. Both of his parents were doctors and he soon found himself studying and interested in medicine as well as boxing. That seems like a pretty good combination, as it might be good to know how to stop bleeding if you spend a lot of time getting punched in the face.

Jeremy joined the military and was working at a hospital in Iraq when his father died. He had been planning on going back to Ireland to work with him at the family practice. Without his father, he decided to move to America and study the field of heart medicine, known as cardiology.

After his day at the beach, he took his new idea to his friend Dr. William Ganz and told him about it. Together, they invented the Swan-Ganz catheter. What the heck is that you ask? I'll tell you.

A catheter is a soft, hollow tube that is inserted into your body. In this case, it is inserted into your heart if you are having heart problems. The catheter helps diagnose what kind of problems you might be having and where. So what did a sailboat have to do with anything?

Jeremy's idea was to put a little tiny balloon on the end of a catheter so that the blood would carry the balloon through the bloodstream to where it needed to go. This meant that a surgeon didn't have to shove it along and hope that it worked out okay. The balloon acted as a sail and allowed the catheter to get pushed by the blood to wherever it needed to go quite easily.

Then, when the catheter was in place, the surgeon would deflate the balloon which would allow the blood to pass through the arteries and veins normally with the catheter exactly where it needed to be. This was a real game changer and this exact device is still used today in most every hospital in the country.

This innovation allows doctors in emergency rooms to quickly know what is going on in a patient's heart or lungs. The fact that it came from watching a sailboat on a lazy day at the beach makes it all the more interesting. What great ideas might you get at your next beach vacation? Great ideas can come from anywhere at any time. You never know!

Are You for Real?

You are hiking alone in the woods in the northwestern part of North America. The wind whistles softly through the trees, but the rest of the world is silent. Suddenly, you hear the crack of a tree behind you and the sound of leaves crunching under feet other than your own. Slightly spooked, you look around. Is that something big and hairy running away from you? You blink and it's gone. What could that have been?

Well, if you are one of the many people who believe in cryptids, you might have just seen Bigfoot!

Here's another possibility... You are rowing your boat on the quiet lake of Loch Ness in the Scottish highlands. Gentle breezes blow the rich green grass on the rolling shores. A soft ripple caresses over the otherwise still lake as you sink your oars into the water and pull the boat forward. You pause, taking in the beauty of the day when you see a small head extend on a lengthy neck above the water. You do a double take, but all that remains is a ripple on the water...

That one would be Nessie, otherwise known as the Loch Ness Monster.

So what do Bigfoot and Nessie have in common? They are both considered cryptids, and are studied by cryptozoologists. By definition, a cryptid is an animal that has not been seen or described by mainstream science. And yet there are cryptozoologists who study these creatures and try to prove their existence.

Cryptids are usually based on folklore and rumor and not factual scientific evidence. While some people make claims that they have seen them, there is no substantial photographic evidence that exists. There is also no evidence that the animals themselves have left behind.

With an animal as big as Bigfoot, you would expect to see some pretty large footprints. Scientists think if there really was a Bigfoot, he would be fairly easy for a professional tracker to follow. They also believe there would be DNA evidence left behind in the form of hair, poop, or the remains from deceased relatives. Lastly, there doesn't seem to be any real pinpointed location where Bigfoot exists. The Bigfoot Field Research Organization reports Bigfoot sightings in 49 states, with Hawaii being the only exception. It seems unlikely that an animal could be that prevalent, and yet that hidden at the same time.

As for the realities of Nessie, scientists are debunking some of those mysteries as well. They have searched the lake and surrounding areas looking for big bones from the Loch Ness monster or its ancestors but have yet to find anything. Many scientists feel that the lake Nessie lives in, Loch Ness, isn't even a big enough ecosystem to support a massive animal such as Nessie. Plus, Nessie is usually thought to be a large reptile, and Loch Ness is way too cold for a reptile to thrive.

All of these scientific reasons lead most scientists to believe that the cryptids just aren't real. But here's the fun thing. In interviews, most of the scientists state that they wish Nessie and Bigfoot were real. They think it would be fascinating to study them and learn about these elusive creatures. It turns out that even scientists think these stories are fun, even if there isn't any evidence to prove it.

How Smart Are You? Well, Let's Measure...

Imagine going to school and your intelligence isn't measured by how smart you are, but rather how big your head is. At one time, people with big heads were thought to be super intelligent, while people with smaller heads were thought to be less smart.

People truly believed that the larger the brain, the more intelligent a person must be. And since these big brains needed some place to live, the larger the skull, the bigger the brain, well...the smarter the person.

But somebody figured out that this just couldn't be right. Alice Lee was a young mathematician and she wanted to disprove this theory. The main reason she disagreed? Most women naturally have smaller heads than men. And at a time when women were trying to establish their rights to have higher education, the bigger skull, bigger brain theory (called craniology) wasn't helping their cause. Men who wanted to keep women out of education often used this theory to persuade universities that women were inferior.

How were brains and skulls measured back then? Well, it was certainly no precise science. Almost all studies were done on dead people's skulls. They were sometimes filled with sand, rice, or some other filler and then the volume used to fill it was measured. You can see that this yielded some very differing results. Yet, nobody seemed to have a better idea.

Luckily, smart Alice did. At a time when most women weren't allowed to go to college, Alice was one of the very first women to graduate from the University of London. She was working with Karl Pearson to get her advanced degree and she really enjoyed statistics and evolution. Karl encouraged her to build on her idea that women were just as smart as men and that cranial capacity (skull size) didn't affect intelligence.

Alice found a way to measure the volume of the skull by taking measurements on the outside of the head. Therefore she was one of the first to be able to measure the skull capacity of living people. Luckily, a group of men from the Anthropological Society volunteered to have their heads measured. They probably figured that their volunteering would help prove once and for all that men were smarter than women. Boy were they wrong!

Alice took her measurements and produced her calculations. The results were all over the place. Men who were considered highly intelligent had smaller heads than their counterparts. In fact, Julius Kollmann, who Alice said was "one of the ablest living anthropologists" had the smallest head of all!

She repeated her study on men at University College and women at Bedford College. Again, results showed no correlation between intelligence and skull size. In fact, some women had larger heads than men who were considered extremely intelligent. These men had to either admit they were wrong about craniology, or that they weren't as smart as the women. I'm sure you can guess what they chose!

Alice Lee was a revolutionary mathematician who paved the way for further research discrediting craniology. And it's a good thing, too! No matter what size head you have, you can be as smart as you want to be.

Help the War Effort! With Silly Putty?

You might be thinking, "How could a soldier use Silly Putty in battle?" The answer to that is only if they were bored and wanted to play and be silly, which is probably not a good idea during a war. Silly Putty was actually created as a complete accident. And it was discovered by scientists who were trying to help the United States during WWII.

The U.S. War Production Board was paying scientists at the General Electric company to try and come up with a good substitute for rubber. This was because Japan had disrupted rubber production all around the Pacific. This caused a rubber shortage around the world and rubber was used to make all kinds of stuff. Rubber was in everything from tires to boots. This made it important to the US army as well as pretty much everyone who drove a car, rode a bike, or wore shoes.

Credit for the invention of Silly Putty is usually given to the General Electric chemist, James Wright. In the lab, he combined silicone oil with boric acid and got something really cool. It bounced and stretched. It didn't melt easily and wouldn't get moldy. But unfortunately for the army, you couldn't make tires out of it. It just wasn't strong enough to replace rubber.

Six years later, a toy store owner saw people playing with it at a party. People seemed to enjoy it so she thought she would sell it from her shop. She called her advertising consultant, Peter Hodsgon, and the two started to sell it. It did okay, but the store owner decided to stop selling it after a while. But Peter thought that it still had lots of potential.

Peter was in debt which meant he owed a lot of money to the bank. He borrowed $147 to get a big batch of the nutty putty made, renamed it Silly Putty and had the idea to put it in little eggs since it was around Easter.

Peter must have thought that he really messed up because he hardly sold any in the beginning. But he got his big break when it was mentioned by an important magazine, the New Yorker. Once the magazine came out, he sold 250,000 eggs in 3 days!

After 30 years of selling Silly Putty, he had sold a whopping 300 million Silly Putty eggs. The $147 dollars he spent in 1949 was now worth $140 million dollars. Wow!

Astronauts have taken Silly Putty into space and nearly every kid everywhere has rolled it around in their hands. Once, a college student dropped a 100 pound ball of Silly Putty off a 3 story roof to see what would happen. (They made sure nobody was walking below.) What do you think happened? The huge ball bounced 8 feet into the air. When it hit the second time it shattered into a million pieces. I would not have guessed that! Now you can look up videos of people dropping it off of really tall buildings, but I just like to have a little bit to play with.

Every Important Innovation Is Usually Met with Really Silly Skepticism

Can I give you a quick pat on the back? I'm so glad that you have the spark of curiosity within you. It is an incredibly valuable trait for anyone to have. An open mind is quite useful to have as well. This story is about how many people have not been able to have an open mind about most of humanity's greatest innovations. It can be really funny looking back at how wrong people were and how silly they were to freak out about new technology. But if you're aware of this tendency that we all have, you'll be able to recognize it when you see it. Let's dive in!

Have you heard the expression, "It's the greatest thing since sliced bread?" This is a funny phrase that means something is really wonderful. But you might be surprised to find out that some people thought that even sliced bread was scary and terrible.

Sliced bread was first sold in 1928. That may seem like a long time ago but your great grandparents were probably living their lives in 1928. That means that sliced bread is actually a pretty new thing compared to how long humans have been alive on the earth. There are newspaper reports from 1929 of people complaining that people were too lazy to cut their own bread and that this was a sign that the whole world was getting worse. Later, selling sliced bread in grocery stores was actually banned for a short time during World War II. Who knew that even sliced bread could be so controversial!

That seems pretty silly today, but I start with sliced bread to show that anything new is often met with confusion, skepticism, and even anger. Humans are absolutely wonderful and amazing, but we can also be very, VERY silly.

Here is a short list of things that seemed dangerous to people when they were brand new: TV, radio, cars, books, bicycles, telephones, airplanes, elevators, railroads, electricity, photography, computers, crossword puzzles, anesthesia, the printing press, the top hat, ATMs, jazz music, pants, and even reclining chairs. Wow! If people were really scared of reclining chairs, that means that pretty much every new thing introduced to humans has freaked us out.

How about the very first telephone? Could people really not want that? When Alexander Graham Bell, the inventor of the telephone, tried to sell the patent to a telegraph company, he got this response.

"The idea of installing 'telephones' in every city is idiotic...Why would any person want to use this ungainly and impractical device when he can send a messenger to the telegraph office and have a clear written message sent to any large city in the US? This 'telephone' has too many shortcomings to be seriously considered as a means of communication. The device is inherently of no value to us."

People either seemed to think that nobody really wanted to talk to people they couldn't see, or that once we had telephones nobody would ever go outside again.

Later, people would have similar concerns about cell phones. One expert said in 1981,"But who, today, will say I'm going to ditch the wires in my house and carry the phone around?" It can be hard to imagine that things will ever be different from what we are used to.

We are always used to our current technology. This often means that anything new and better, might seem unneeded. Even cars! Lots of people thought that the invention of the automobile was ridiculous and that nobody would buy a car because everyone already had a horse.

People thought that electricity would make us lazy and that oil lamps were just fine. They weren't, of course. They were much more dangerous and harder to use. Brought on by human skepticism, the guy who invented and wore a top hat in public for the first time back in 1797, actually caused a riot and was charged with a crime. According to police, his new style of hat caused women to faint, children to scream, and dogs to yelp. I would have LOVED to have been there to see that!

A highly respected scientist once said that the modern world was full of too much information. It overwhelmed regular people and was harmful to people's brains. This modern world that he was talking about was the world in 1565! Conrad Gessner was the scientist who said that and he thought the printing press was really dangerous. This seems especially strange because he had published over 70 books when he said this. He thought that ordinary people couldn't handle access to so many books. I wonder what he would think of you reading this book?

There was another guy who didn't like the printing press. He was a leading scholar of his time who said that mechanical printing was morally inferior to handwriting and felt that all books should be copied by hand. He probably wouldn't like this book very much either.

Even earlier, (2,400 years ago) way before the printing press, there was a guy who was even against... writing things down! The funny thing is that he is still considered today to be one of the smartest people to ever live. That man was Socrates, the philosopher in ancient Greece. This goes to show you that even really smart people can think some not-so-smart things. Socrates felt that children could not tell what's real from what is made up and that if they wrote things down, they wouldn't be exercising their memories. So he REALLY wouldn't have liked kids reading something. There's another guy who would leave this book a bad review.

So in everything from new technology, to food, to fashion, we tend to be scared of new things, inventions, and innovations. It's good to remember this as it might help you to keep an open mind. This will make you smarter than people who are unnecessarily fearful. It's also fun to know that a lot of history's most famous philosophers would think that a kid like you having this book is a dangerous thing!

YOUR REVIEW

What if I told you that just one minute out of your life could bring joy and jubilation to everyone working at a kids book company?

What am I yapping about? I'm talking about leaving this book a review.

I promise you, we take them **VERY seriously**.

Don't believe me?

Each time right after someone just like you leaves this book a review, a little siren goes off right here in our office. And when it does we all pump our fists with pure happiness.

A disco ball pops out of the ceiling, flashing lights come on... it's party time!

Roger, our marketing guy, always and I mean always, starts flossing like a crazy person and keeps it up for awhile. He's pretty good at it. (It's a silly dance he does, not cleaning his teeth)

Sarah, our office manager, runs outside and gives everyone up and down the street high fives. She's always out of breath when she comes back but it's worth it!

Our editors work up in the loft and when they hear the review siren, they all jump into the swirly slide and ride down into a giant pit of marshmallows where they roll around and make marshmallow angels. (It's a little weird, but tons of fun)

So reviews are a pretty big deal for us.

It means a lot and helps others just like you who also might enjoy this book, find it too.

You're the best!
From all of us goofballs at Big Dreams Kids Books

Made in the USA
Monee, IL
22 September 2022